ONE WAY … SEARCH

the A–Z …

ONE WAY TO DO RESEARCH

the A–Z for those who must

James Calnan, F.R.C.P., F.R.C.S.,
Professor of Plastic and Reconstructive Surgery,
University of London,
Director
Experimental Plastic Surgery Unit.

William Heinemann Medical Books Ltd
London

First published 1976

ISBN 0 433 05012 8

Photoset, printed and bound
in Great Britain by
REDWOOD BURN LIMITED
Trowbridge & Esher

CONTENTS

ABOUT THIS BOOK

Research is like completing a jig-saw puzzle. Every piece looks similar yet different. Every piece carries equal weight and all have to be fitted together to make good the whole. In a way, of course, that is the trouble; the missing piece (whether it be plan, hypothesis, ideas, people, resources, management, money, or just plain good sense) suddenly becomes the most important. Yet to complete the jig-saw one must have an idea of the picture expected. Hence, this book. It does not tell you how to do research but tries to indicate in advance just what the whole picture may involve. There are no smooth-running chapters because research is not like that; the slow, thoughtful arranging of a jig-saw puzzle gives a far better idea of how things really happen.

Many people want to do research, or at least they say they do, but have no idea how to go about it. When they consult a library for help there is virtually nothing written and when they ask their superiors they are told that research cannot be taught. Is this entirely true? It is a widely held belief. It has always been assumed that research means doing something, but what of the principles or are there none, as some would have us believe? Certainly, there are some aspects of research which can only be taught on the job (commonly called experience) and anybody who has had the slightest success in research, knows that you learn by doing. So what can a book do for you? It can enlarge your ideas about the subject in a general way; it can help you to understand what is required; it may help you to find you what *not* to do (a famous precedent is The Ten Commandments). You may not have the ability to do research, so a book can warn you off although that is not the object of this one.

It really is a curious situation where some men have intense curiosity to find out more about the universe, while the majority are quite content with what they see and do. Yet, we are all told that advancement in knowledge can only come by research. If this is true, who is to do it? A devoted band of intellectuals or anyone who can take the trouble to think and ask and do? The evidence is strongly in favour of the latter. They may never become Nobel Prize winners unless of outstanding merit; the vast majority of research is less world-shattering, but important all the same. The best way to do research is to do research, but the moment we decide to do research most of us are faced with the problem of knowing where and how to begin. This book only points you in the right direction – no more.

It was not written for the expert and is in no way a book on "How to . . .", the aim has been much less pretentious, a modest "One way to . . .". It is no cookery book of recipes for the novice to follow but a serious attempt to explain what has to be known and understood before you start and long after all is finished. Some never learn because no one ever told them. The book is not for the expert who will have acquired such information at the time when he was engaged actively in research. Rather, it tries to set out what so many experienced people know yet fail to pass on to others who are just beginning; not from any malicious intent to see them flounder, but simply because a great deal of that knowledge was never presented in a formal way when they first learnt it; it is the knowledge of experience and has become almost intuitive.

On the contrary, this book is as much for the physiotherapist who wonders if her work really benefits the patient (or would he get better anyway?); for the nurse who objects to being told that "This is the way we do it at St. Wormwood's", yet her own observations suggest that tradition is quite wrong; for the speech therapist who suspects that lessons in articulation to those with gross nasal escape are quite useless, but does not know how to set about proving it; for the radiographer who is conscious of the cost of a film and wishes to reduce the number taken annually, yet does not know how to set about costing the casual patient; for the mundane practical doctor who does not aspire to be called "academic" but who would like an answer to a question which has bothered him in practice for years; for the technician who would like to do his "thing" on his own and stand or fall by what he finds; for the pharmacist who suspects that placebos are better than some drugs on the shelf; as for the medical administrator and for all those allied to medicine who can ask,

"Why? How? Who?" and mean it.

The views expressed are not necessarily those of my colleagues. I have watched them, admired them, envied them, despised them, and then unashamedly creamed off the best to learn a great deal. The mistakes, of course, are entirely my own.

With no disrespect to the fair sex, the male pronoun has been used throughout (except on a few deliberate occasions) because to use both continually is clumsy in English. It is realised only too well that most of what is written applies equally to both sexes. In the same way there is no index because the material has been arranged in alphabetical order. It is not a dictionary, however, so that if you do not find the title you want, try another, which at least will make you think – and that is to do with research too.

ABSTRACT

The purpose of an abstract or summary is to enable busy people to get the gist of a report or published paper without having to read it all. It allows one to keep abreast of new work without having to wade through pages of tasteless detail. Writing an abstract is not easy: it requires particular skill to give only the essentials that the cursory reader requires.

An abstract should be limited to about 400 words, say one page of typing on A4 paper, and will incorporate:

- The purpose and scope of the work
- What was done, how it was done.
- The findings.
- Conclusions and their validity in reasonably clear and concise language (itemised if necessary)
- Recommendations for tomorrow.

Many journals now publish an abstract of a paper at the beginning (to induce the reader to look further) and a far larger number include abstracts as the sole offering. Indeed, there is mounting opinion that information is best disseminated by abstracts, the author being requested to supply full data on demand to the incredulous and the interested. The idea has great economical appeal, but a wonderful opportunity for unscrupulous and inexact brethren.

ACKNOWLEDGEMENTS

One would have thought it easy to say "thank you" when someone has done you a service, yet most people find it a difficult thing to say. Who should be thanked?

Throughout, I have tried to emphasise the importance of consulting an experienced researcher for help. Ask your chief, if he has written several successful papers: do not consult him if he has not. If he helps by correcting the script and has materially aided its final form, this should be recognised in print. As Montagne noted, "Gratitude is a hope of further favours", but to curry favour by acknowledging a senior when he has done little is unworthy, for flattery corrupts both receiver and giver.

Technical help in experimentation should always be recognised and with it any financial help from a grant-giving body. The artist and the photographer generally do not sign their own work, so it is the duty of the author to give them credit.

The acknowledgements are usually placed at the end of the paper. A simple statement is all that is required; it should be short and factual, not long and flowery. But it must be done! If your helper has done so much that he really should be an author, whatever his status, then think again; better still, put his name at the top.

ADVANTAGES OF RESEARCH

The advantages of research to society are obvious to most people; just look around and see how newly acquired knowledge has been put to use by technology. But, for good or evil? This is the concern of many today.

Gerald Leach used the term "The Biocrats" to describe those working in areas of research where successful experiments may influence society greatly – population control, test-tube babies, the eradication of congenital deformities, transplants. His attitude assumes that every research worker adjusts his experiment to his social conscience. Not so. Good research follows a recognised pattern. Society can demand research in specific areas where knowledge is needed: whether it is obtained is quite another matter. Professor Peart had the same idea in his essay, "Should research be left to the researchers?" and his answer was "No". It is true that in times of national crisis the abilities of all research personnel (I think that is the correct word) are mobilised with a single objective in mind, to defeat the enemy. Hence research becomes war-orientated, partly from patriotism, partly from emotion, partly because that is the most pressing need for self-preservation. This is not so in peacetime where the purely intellectual acquisition of knowledge is far more exciting.

In Britain the Dainton Committee advised the status quo, but Lord Rothchild wanted a customer-contractor principle with the old idea that "he who pays the piper calls the tune"; old fashioned because it fails to recognise the devices of the piper. Money will not produce research results; oh yes, money will be accepted and put to good use (not always to that intended by the donor) but if the individual does not have the burning desire to succeed in that area, results will not come. Research demands inspiration and a great deal of freedom for creative thought. Leach writes about cost-effectiveness, an important consideration in anyone's economy but almost impossible to apply in research. What is being costed? The intellectual capacity of the individual?

Gordon McLachlan (*Lancet* 1967 *1*, 629–631) thought that to secure vast public support for research and for the latest equipment and techniques for saving and maintaining life is relatively easy. On the other hand, the organisers of medical care on which the effective application of scientific knowledge depends "have a hard row to hoe". Maybe. But remember that, on the whole, the scientific method produces results and management is still in its infancy without a method of measurement or accurate means for comparisons. It is indeed true that the scientific method cannot yet be applied to those things that cannot be measured. In any case, many current problems are social ills, not medical diseases, and so require social reforms, not tranquillisers.

The value of research for the individual is clear cut:

- It provides him with a view of science which he otherwise might not obtain or could only acquire second-hand by extensive reading of other people's opinions. You may think yourself "King" of your patch, but it is only by seeing the achievements of others that a true perspective is obtained. In fact you are a pigmy (but may have an important part to play, nonetheless).
- It widens his horizon of a subject, in medicine generally, in the interactions of health and social conditions.
- It allows a person to concentrate on a single thought, to the exclusion of all others, at least for a time.
- A condition can be studied in depth. William Harvey said: "To search and study the secrets of Nature by way of experiment". He did.
- It forces one to search and read the relevant literature with discernment and appreciation. Reviewing earlier publications on a subject and then setting up a valid experiment gives a valuable insight into the problem of supporting an hypothesis and a better understanding of other people's work.
- Personal skills and thinking are always improved and these are carried over into clinical practice to everyone's advantage.
- The discipline required (and acquired) to carry out research successfully is something well worth the effort; it is never lost. An opportunity to think and act as an individual, to raise a lone

ADVICE

Advice comes in two sizes, that large economy pack and the small travelling sachet – just like detergents. The former is advice asked for, the latter advice given.

When people ask for advice it often means that they really want someone to talk to, someone to listen (Somerset Maugham said "they only want praise" – a cynical view). When you are prepared to listen, people will tell you what is in their hearts; but listening requires complete attention, silence and interest. Perhaps, before you've made a single suggestion, the troubled person will have left, at peace now that he has seen the way ahead.

We all want to express what is in our hearts; sometimes we need pity, encouragement, help. Listen to anyone who wants to speak to you, but listen with genuine interest, without becoming bored and showing it. Lots of people have never found their true-selves because no-one has had the patience to listen to them with complete attention. Listening to others is an important feature in research and how we get along with them is the key to whether we are good mixers or bad, the test and the answer.

In clinical practice it is a truism that the man who listens to what his patient has to say is a better doctor (and a more accurate diagnostician) than the one who rushes to carry out special tests before hearing the end of the story. In research, one has to learn to listen. As recompense, there may be the germ of an original idea in what the speaker has said; a pearl worth waiting for, even though it may be surrounded by a hard shell.

Giving and taking advice is a two-way affair and two golden rules should be observed:

- Don't ask for advice unless you intend to take action.
- Don't give advice unless asked.

Often, advice in research is a form of mutual aid, an old concept in new terms. One of the great pleasures of working in a large institution is the ease with which one can pick up the telephone to ask for and receive advice. Such temporal help lays a debt which can be repaid in kind at a later date. Nothing changes hands except good will, a rich harvest spelt out 2,000 years ago: "Cast your bread upon the waters without thought of gain and the reward will be manifold".

ANALYSIS

Some people collect facts, perhaps we all do, which represent useless information. Television and radio quiz programmes encourage this occupation and appear to make it respectable. It is not. Useful facts will be of four kinds:

- Statistical calculations based on publications, reports, a pilot trial, and unofficial sources.
- Those known facts which may be directly or indirectly related to the project, or even of no apparent relationship.
- Other facts not directly related to the project but having significant relationship with background data.
- Some facts implied in the working hypothesis which may lack validation or sufficient confirmation, which pose three further problems:

1. the establishment of the relationship of the data to the research projects,
2. the reliability of the data, and,
3. an assessment of conflicting data which has to be confirmed, rejected or reconciled. All speculative data must be discarded immediately and the remainder graded and classified for reliability (the appraisal requires fine judgement).

Most research produces far more information than can be used immediately. It should be stored not abandoned for it may be significant in relation to other facts later. The processing and refining of facts is an act of interpretation. Analysis and interpretation are the true results of any research, not the end product for they should begin the moment the first facts are uncovered (see NOTEBOOK DISCIPLINE) by a constant critical appraisal of incoming information (the alternative is an untidy mass of data in various stages of validation). The redirection of research cannot take place unless analysis is continuous. The facts collected may not all be relevant to the project; these have to be weeded out (but not discarded) if only to reduce the mass of data to manageable proportions. Criticism looks for the evidence, analysis supplies it.

Analysis involves five steps:

- Investigation of bias in the data, which applies particularly where the information is qualitative and not quantitative. It is eminently possible to collect qualitative data which is statistically valid, but bias must be eliminated at the start, by random allocation, and not sorted out at the stage of analysis (see CONTROLLED CLINICAL TRIALS). Bias may be of three kinds; in the source of the sample, the questions and answers and in the analysis of the data.
- Quantitative analysis by editing, coding, tabulating. The editing is the sorting out of the material; coding and classification its categorisation; and tabulation the production of tables and their statistical assessment of useful and relevant information. With good notebook discipline most of the important tables will have been constructed already.
- Qualitative analysis is more difficult because opinions have to be weighed, yet never attain the same reliability as quantitative data.
- Interpretation of the facts, and often this is what most people understand by the word analysis, which the dictionary defines as "the resolution of anything complex into its simple elements" but adds also "the exact determination of its components".

Interpretation is more than that, for it really sorts the wheat from the chaff and tries to make a coherent story. A research project which does not include an interpretation of the findings is incomplete, because interpretation bridges the gap between the facts discovered and their meaning for future action.

- The recommendations either support or refute the original hypothesis on which the research was based. They may raise new questions, some of which can be answered from the work done but generally indicate the need for more and specific data. At this stage the research may be written up for the benefit of others or put aside until the material can be fitted into a more complete and unified whole.

ANIMAL LICENCES

Any experiment on any animal poses certain responsibilities for the actor. The reasons for doing animal experiments are various, from studying a biological system to testing a new hypothesis of the action of a drug or of a surgical operation. In the last event all will have to be translated to Man for the final experiment (Does it work?). In Britain, and other countries, Home Office authorisation is necessary for most experimental work involving animals. Licences for experiments on animals are not issued to allow one to improve manual dexterity nor to improve techniques of established surgical operations, nor is anyone licenced to inflict pain, consciously or unconsciously, on any animal. Sadly, it is often cheaper and possibly easier to perform the experiment on Man in the first place.

The Cruelty to Animals Act of 1876 (old, but still viable) lays down that:

- Animal experiments may be performed only by those who have first obtained the appropriate authority from The Home Office (general application forms are available).
- The experiments may only be carried out in designated premises, registered and inspected by The Home Office.
- The nature of any experiments must be, within certain limits, acceptable to the Home Office.
- Premises and experiments are liable to Home Office inspection without prior warning and experimental records may have to be produced.

All this sounds daunting enough but it is a two-way affair; it protects the experimenter from outside interference but makes him personally responsible for his conduct of experiments. It also involves the institution in the management of its staff. Now to details and, fortunately, the Inspectorate issues "Notes for guidance in completing forms"

The licence form is fairly general in content, covering an outline of the nature of the experiments proposed, where they will be done and by whom, and a note of special certificates submitted. This and other forms (the special certificates) have to be signed by the applicant, the head of the (University) department and the President of the Royal College of Physicians/ Surgeons or the President of the General Medical Council. An interval of 6–12 weeks is common between submission and approval.

Licence alone permits work only on animals that are anaesthetised and will not be allowed to revive consciousness. Completing the various certificates can be a trial. Certificate A is required for minor procedures without anaesthesia, such as altering the animals diet, taking blood or giving an injection. The description of an "experiment" must therefore be written in such terms – the withdrawal of body fluids, exposure to irradiation, needle biopsies or implantations, and not a repeat prescription of the original plan. If procedures severe enough to require an anaesthetic (with subsequent recovery) are envisaged Certificate B must be completed as well. Certificate C is demanded when any experiment will be used to illustrate a lecture (on the whole a foolish gesture except for those with the supreme confidence that all will go well) and the nature of the audience to whom this will be delivered is expected, as well as details of the demonstration. Certificate D, thank goodness, no longer exists.

Certificate E cannot be held alone (there must be A); it must exactly repeat the wording of the linked A and extend it to cats and dogs (remember that the British value their dogs more than their children). Certificate EE is required to extend Certificate B for cats and dogs. Certificate F is required for all experiments on horses, asses and mules (probably applicable to Veterinary Surgeons only because no-one else can afford that luxury) and is linked with Certificates A, or C, or with the Licence alone.

The sections on the various forms require comment. The "Description of the experiments to be performed" should relate to the procedures intended under the particular certificate included with the general licence application form. The latter should set out precisely what will be done (but briefly). Moreover, it must be stated how the experiments are designed to advance man's well-being by new discovery of physiological knowledge, methods of saving life, prolonging life or alleviating suffering. For cats and dogs, the reasons for using such animals and why no other is suitable, have to be described. When the applicant is not a national of the U.K. a form of "undertaking" must be submitted by the supervisor of the proposed experiments.

Applicants must check that after signature by the sponsors the form is indeed fully and properly completed and they are strongly advised to include a letter with their submission, addressed to the local Home Office Inspector, giving more informal details of the background and intentions than can be accommodated on the official application forms.

Licences are normally issued for five years and have to be renewed then but new certificates can be added as the nature of the experiments changes. Changes of name or address should be notified to the Home Office, although correspondence will usually be sent to the institute where he works (if he changes institutes he always has to resubmit the licence for authorisation to work in the new premises). At the end of each calendar year, the licensee has to make a return of the number of animals used and include a copy of any relevant publications.

Copies of the 1876 Act and forms of application may be purchased from Her Majesty's Stationery Office, 49 High Holborn, London, W.C.1. When completed they should be sent to the Home Office, Cruelty to Animals Act Inspectorate, Romney House, Marsham Street, London, S.W.1. with a covering letter. The applicant is also advised to join and support The Research Defence Society, 11 Chandos Street, London, W.1.

A useful paper to consult is "The use of laboratory animals" (Annals of Royal College of Surgeons of England, 52, 193–197 – March, 1973) which contains notes for the guidance of research workers and has a list of related publications as an appendix.

APPENDIX

An appendix is not the place for a whole load of junk; that is, irrelevant or useless information which you may have gathered on the way through some simple research. It is the place where material needed to support the body of a report or paper is inserted so that great detail or volume will not impede easy reading. It is not correct to put there all diagrams, charts, tables, graphs and details of experiments which are required to sustain the main theme.

In a report, for instance, summaries of results obtained at other institutions, important correspondence and special graphs are quite properly matters for appendices; they can be examined by those most interested in details but they are not required for an understanding of what the report was all about. If a complete bibliography of publications has been made, this too will become an appendix, even though the most important references will be in the text and perhaps even the full details noted there.

ARTIST AND ILLUSTRATOR

The term illustrator includes both the artist and photographer. In research both play a very important part in passing on the results of your work to others. Both should be consulted early on, before the paper for publication is composed and preferably at the stage when the data are being sorted out. Ask their advice for:

- the best lay-out of the material. Will it have more impact as graphs, tables, histograms?
- the composition of slides for projection in a lecture. What should be put in, what can profitably be left out?
- new techniques of illustration which may make the work more easily understood by others.

Both will be pleased to be asked to comment and the budding author or lecturer will gain immensely from experts in another field.

AUTHORS

A time will come when the publication reports the work of a team. Who should be an author? This is a more difficult question to answer than at first appears. If it is customary for the name of the senior man to appear on all publications from the department then he must read all, correct and comment; but on the whole, this is little short of piracy.

Alexander (1953) said that the essential requisite for authorship of a scientific paper was the contribution to creative thinking, and asked, "How often is it the product of 6–8 individuals thinking in unison?". The current practice of multiple authorship undoubtedly stems from the growth of research teams. When there are multiple authors, one man only must write the paper. Multiple writers produce a patchwork of different styles, and to allow each to write a section is quite ridiculous!

Creative thinking should not be confused with expert consultation. My own rule is that an author must have contributed something substantial. If he has only corrected the typescript or given advice this can be acknowledged at the end of the paper. In general:

- Authors arranged in alphabetical name order, should all have contributed equally to the work.
- The first name on the list implies that he has done the greatest amount of work and other names are arranged then in proportion.
- If there are only two names, the first has probably done the work and the other has written it.
- If there are only two authors, the senior may elect to go second, even though there has been equal contribution, because the advantage may be greater for the junior. In the next paper, in a continuing study, the senior man can be the first to redress the balance.
- Commonly the senior's name comes last, by convention,

where contributors are equal and multiple.

- If the contribution is likely to be part of a thesis for one of the authors, then his name should be the first whenever possible; he will probably have done most of the work anyway. If this order is not observed it may jeopardise the validity of the thesis as being essentially a personal project.

voice as opposed to that of a group, to fight conformity and walk the other way instead of following the crowd, to have a novel idea and show that it might be true all by yourself; all these provide self-satisfaction and intellectual fulfilment. Techniques developed in other disciplines, in biochemistry for instance, may be adapted to the welfare of your own patient.

● Research makes a better man. As Bernard Leach, a famous potter in his time, once said: "Every improvement in the things that men do is followed rapidly and directly by improvements in the men that do it". This is certainly true because research is a matter of the spirit and often little to do with physical dexterity, even though such may help.

Research ignores departmental and geographical areas, specialties and seniority. In becoming a researcher, in joining the club, you will have access to a permanent source of informed opinion. Moreover, it is now possible to bridge the gap between the academic and research committee on the one side with the world of practical affairs on the other.

The purpose of research is to:

1. Increase our knowledge of the natural world
2. Improve techniques in daily practice
3. Organise work efficiently and productively
4. Apply knowledge to particular problems and individual disciplines for the common good.

These, surely, are all worthy subjects; they indicate the importance of research for the individual and society. None of us can stand still in life; we either go forward or slip back. Moreover, improved technology may save time, money and effort for others. The thrill of discovery is a very personal reward, to be experienced to be understood. To have to learn to formulate a problem, to nurture it, to study it in an experimental set-up, to arrive at a conclusion, requires clear thinking and is a discipline in itself; in order to learn you may have to do it all on your own. That is largely what research is about.

AVERAGE

There are three main types of average – the mean, the mode and the median; and the first type may be subdivided into arithmetic mean, geometric mean and harmonic mean. These are all different types of representative value and, being recognised as such, the noun "average" is rarely used in statistical work when referring to actual values. The statistician, when mentioning these values will always refer to them by their specific names. If he wishes to talk about them collectively he will refer to them as measures of central tendency rather than as averages. But elsewhere, they are often referred to as averages without any indication of the type of measure being used and, since each type is derived quite independently of the others and has quite a different meaning, it is not surprising that this sometimes causes misunderstanding.

So speak of "the average" with caution, and explanation.

BEST OF BOTH WORLDS

If you live in a lab. you become known as a lab. man; if you operate on dogs, rabbits, mice, rats you are a surgeon of these only. The labels are generally applied by colleagues and others who may not appreciate what you are trying to achieve, even may not wish to know. A great pity, because you may have the makings of a great surgeon or physician. How to shed these illusions? Simple. Continue with clinical practice and let your patients know about your research interests; they, the patients are far more discerning than you might imagine, and once a patient-doctor relationship is established, it is difficult to break unless of course, you are a tactless, unimaginative fool who cannot discern truth from falsehood. Compassion, interest, direction of purpose all have their reward, but the ultimate reward is the relief of human suffering. Patients are human, understanding, helpful and co-operative – if, and only if, they are told what it is all about.

COMMITTEES

Whether you like it or not, sooner or later you will be asked to serve on a committee. For the first time in your life the views you have to express may not wholly be your own.

Committees generally are a bore; certainly by reputation, if not always in practice. Yet this should not be so. Committees are formed to conduct business: if well done, every member is satisfied and the meeting has been a stimulating experience, to hear the views of others, to listen to well-organised arguments, to watch how your colleagues react to new situations. If done badly, a general air of apathy and frustration will pervade and the worth of the committee will be questioned.

Every member has a duty to those whom he represents and to the Committee as a formal body. So ask yourself three questions: Why am I here? What do I have to do? How can I do it? If you know and understand the answers to these you have little to worry about. Doing well in committees is an acquired skill, to be learnt like any other skill by watching and doing. Watch the technique of a good Chairman, the nuances, the tact, the clarifying sentences by which business gets done with efficiency and fairness. Committees may be formal or informal. On the whole, it is better to have a formal committee meeting and then allow informality to creep in. Planned informality often leads to time-wasting talk and no action; waffling carries no weight.

THE AGENDA

Every committee meeting requires an Agenda; it should be set out like a menu – short and appetising. However brief or limited the meeting is expected to be, without a written agenda things will get out of hand. Indeed, the secret of successful meetings is to have an agenda, stick to it, allow only one person to speak at a time and to be brief. Every agenda sheet must include the time, date and place of the meeting. An example might be as follows:

- Chairman's remarks: these can be as brief as "Good Evening", but should also introduce any new members by name (and a little about them if appropriate).
- Apologies from members who are unable to attend.
- The minutes of the last meeting, if recorded, are either read aloud or, if circulated previously, then signed by the Chairman as a correct record.
- Matters outstanding from the last meeting or arising from the minutes.
- Correspondence, either from members or others, to the committee.
- The main business. Here one can include reports on finance, membership, other activities and special reports from subcommittees. Reports of Government documents or from delegates to other meetings are also included.
- Any other business: this section is for minor items and such trivia as members may wish to discuss openly, certainly not for main business items.
- Date of the next meeting.

An unwritten law of all committees is that the time taken over each piece of business at a meeting increases in inverse proportion to the number of items on the agenda. The skilful secretary, therefore, inflates the number knowing that an agenda with ten items will be covered more speedily than one with only five.

All these items can be listed more briefly on an octavo sheet and, indeed, may be of a permanent form:

1. Chairman
2. Minutes
3. Correspondence
4. Reports
5. Unfinished business
6. New business
7. Delegation (subcommittees, representatives)
8. Next meeting

RULES OF PROCEDURE

For formal meetings there is usually a set of rules of procedure. These vary between committees according to the nature and importance of the work undertaken. For regular, more informal meetings, a useful outline for members is as follows:

1. Our purpose is to pass on information and resolve problems.
2. The meetings will take place at the same time, the same day and the same place, regularly (like the T.V. news) – every week, month or quarter.
3. We start at 5 p.m. and finish at 6 p.m., no matter who is absent.
4. No attendance taken, just sign the book.
5. No notice of meetings will be sent in advance.
6. There is no stigma attached to non-attendance.
7. When a member fails to attend three consecutive meetings, without written explanation, he will be presumed to have resigned.
8. Minutes, of one page only, will be circulated within one week of the meeting.
9. Sub-committees will deal with details of business and report to the next meeting.
10. Objectives will be defined and discussed at quarterly (or yearly) intervals.

Such instruction may seem a little hard, but in industry where the executives are responsible for the day to day survival of the company, they seem to work well.

DECISION MAKING

The purpose of every committee is to assemble facts and opinions, and then to come to a conclusion. The conclusion is usually the resolution, which requires a proposer and seconder. A vote may be taken to decide whether the resolution is carried or lost. Sometimes, there is an amendment to the original proposal. Of course, if there is no seconder, the original proposition or the amendment fails and is not discussed further.

When a resolution is put in clear, precise terms, and a seconder has been found, the Chairman will then allow debate. When all discussion is over, the proposer should be allowed the final word – to refute arguments, to answer questions and to briefly restate the proposition. He should not be allowed to re-open the discussion by advancing new information, appealing to emotions, or denigrating previous speakers.

Modern convention favours agreement by consensus opinion rather than by the more formal voting. In such cases the Chairman senses the wishes of members and sums up in his own words a suitable resolution. On these less formal occasions the wording can of course be challenged, for the Chairman may have misinterpreted the mood of the meeting.

Decision making requires answers to these 8 questions:

- Is it our responsibility to decide?
- Do we understand the problem?
- Have we asked the right questions?
- Is it feasible?
- Is the timing right?
- What are the alternatives?
- What are the risks?
- What are the likely consequences?

An unsatisfactory answer to any one of these should preclude trying to move on to the next. Most bad decisions are made because insufficient information is available at the time; the correct course is to postpone any decision until it has been acquired. Committees make cowards of us all because, when a decision is staring you in the face and can be made by one man, so often members suddenly become anonymous and funk it.

MINUTES

Not every meeting will have a secretary to record the minutes of the meeting, nor wish to do so. The minutes are a record of business transacted, not a record of what each individual said. Indeed, sometimes a member will state that what he is going to say should *not* be recorded: this is rather pretentious and shows that the member does not really understand the purpose of minutes. There are advantages in having minutes and in circulating them to members before the next meeting; they can be brief and serve as a permanent record of the subjects dealt with. The minutes, like the agenda, should always report the date, time and place of the meeting, who was Chairman, apologies for absence and the business in the order in which it was carried out.

A committee meeting which commonly lasts more than one hour is either inefficient or meets too infrequently. Parkinson's Law applies to committees as much as it does to research, or any other activity; it must be reversed and certainly resisted. Some people rapidly acquire the reputation of "a good committee man", but this is unnecessary. It is usually more important to get your own way than agree with all that is said later.

COMMON SENSE

An over-worked word, but rather important. What is common sense? The dictionary does not define it clearly and the exact meaning has changed over the centuries: from "ordinary, normal, or average understanding" (without which a man is foolish) to "an internal sense which was regarded as the common bond between the five senses" (*O.E.D.*). In the north, they call it "nous" – a kind of innate judgement. Certainly, some have it, others don't.

Yet it is a valuable asset in research, if only to realise when to stop. Surgeons all know of others whose main claim to fame is the ability to make a simple operation difficult. Few follow suit because common sense warns of trouble ahead.

COMMUNICATION

Telling other people what to do is often called "communication" but it usually isn't because they

don't listen,
don't understand,
resent it,
can't or won't do it.

So, what really is communication? Even Drucker (1972) finds it hard to define; it is not the same as information. He suggested that there were four fundamentals in communication:

- perception
- expectation
- involvement
- information, which presupposes functioning communication.

Perception implies that the recipient of information must listen, must have had some previous experience and also be capable of receiving the message (which requires that the terms and words used are familiar). We see and hear what we expect to see and hear; the unexpected is either not received, ignored, or misunderstood. Involvement concerns emotions and beliefs. The credibility of the message to be put across, that is, the propaganda of communication, must therefore appeal to both; hence communication is more effective, the more levels of meaning it has and the less possible it becomes to quantify. Information on the other hand is specific, logical, and economical of data; the fewer the data usually the better the information.

In research and in industry, utterances travel downwards but communication occurs upwards. The most junior member of the team must be allowed to pass his information to the top executive level and not just be a receiver of messages. Fletcher (1973) pointed out that good communication is difficult and demands a great deal of thought, time and effort. He listed five principles of good communication, which apply to research:

1. The purpose is to effect a change in the recipient; if it does not, it has failed.
2. The change in the recipient must be seen to be advantageous for the individual; if not, the message will be ignored.
3. The communication must be in terms that are matched by the knowledge of the recipient (see Drucker's similar comment earlier).
4. Communication is conveyed not just by words, but by attitudes, gestures, actions.
5. The value of any communication is judged by its efficiency and this means that there must be some form of supply of information in the reverse direction from the recipient ('feed-back').

There are 3 main forms of communication in research: the spoken word, the written word and illustrations. The style and format of illustrations will vary according to the subject and the occasion; if a difficult subject is being talked about, then several illustrations may be necessary to explain it whereas in a published paper the number of illustrations will be limited and their complexity increased because they can be studied at leisure.

Many people fail to realise that:

- English is two languages, spoken and written, and that the difference between them is increasing all the time. The easiest way to appreciate this fact is to secrete a tape recorder when conversing with colleagues and later to read aloud a passage from a book. The former will immediately be recognised as spontaneous conversation, even though it may not make sense on "playing back" (but it did at the time) and the latter sounds just what it is, a public reading. Here is a brief and incomplete list of the differences between the spoken and written language.

Speaking	*Writing*
Grammar less obvious	Grammar important and may be either good/bad/ correct/incorrect, word order gives the clue to meaning

Pauses not always equivalent to punctuation	Pauses are punctuation
Literal meaning can be reversed	Meaning of words is according to dictionary
Sentences may be incomplete	Sentences must be complete units
No capitals in speech	Capitals start new sentences
Can point to illustrations and say "look at that!"	Impossible in writing where "see figure x" is correct form
Can use "excess" illustrations to bring out opinions, ensure continuity of speech, add allusions	Severely limited by editor; illustrations supplement or substitute the written word
Humour valuable	Humour difficult and, in general, frowned upon by editors
Alterations of pitch and tone can alert the audience to a change of subject	Impossible in written language
Pronounciation	Impossible in written form
Sounds can be used to good effect—voices of patients, music—and they can be dramatic	Irrelevant in writing and drama not welcomed
Personal pronouns used often	Rarely justified because medical writing tends to be impersonal
Meanings can be questioned	Not possible because confusing
Emphasis by repetition and accentuation	More difficult in writing, except by the use of italics or devious repetition
Word distinction is blurred unless context established: "The sun's rays meet" "The sons raise meat"	Improbable in written language unless jargon introduced
Facial expression can convey disbelief, reverence, amazement, anger, grief	Much more cumbersome in written language

● The lecture, whether formal or relatively informal, and the paper for publication have to be composed in the same way that a composer has to write his music.

The aim of the lecture is to:
entertain
stimulate
inform
teach and persuade the audience.

Moreover, this is the correct order of priorities; for the written paper the order is reversed. For the written paper the audience is largely unknown, but for the lecture one always has some idea who they will be and so you must try to answer these 6 questions before planning a single word:

1. What will amuse them?
2. What will be of particular interest?
3. How much can they absorb?
4. What do they need to know?
5. What do they want to know?
6. What makes them tick?

● Spoken and written communications have two quite different structures:

Lecture (Total subject approach) — *Written paper* (Composed step-by-step)

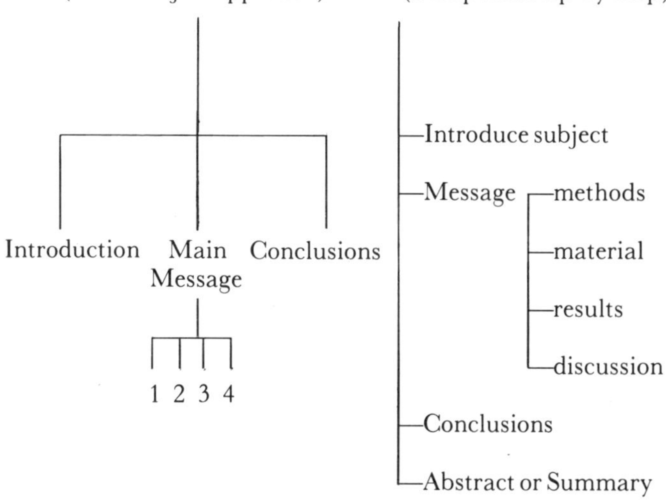

The lecture is planned and delivered as a single offering. Its structure is simple; the attention of the audience must be captured by the first sentence uttered and then held until the end. Too many people worry about what to say but forget that they have been invited to speak because it is recognised that they are experts in the subject or have something important to say. The composition of the main message of a lecture, then, is usually a matter of selection and compression. Because you have something to say you were invited, so why not settle for four aspects of your subject:

What's new in research?
What are the clinical applications?
What are the difficulties and unwanted effects?
How's the future?

Hence you should

emphasise the subject
present it in the most effective form
put it together neatly (you will know too much anyway and have to do some radical surgery to make your contribution palatable).

● The amount of time spent in presentation and preparation is vastly different in a lecture.

	Delivery Time	*Preparation Time*
Introduction	20%	80%
Main Message	75%	10% (because you know it and, anyway, that's why you were asked to speak)
Conclusions	5% (the conclusions should be self-evident)	10%

The paper for publication requires about 10 separate headings and the preparation time varies according to the complexity and detail required for each. Usually the "results" demand detail and the "discussion" a closely argued case. In a lecture both are out of place.

COMPUTERS

The best computer is the human brain because it can accept data and give judgement. The machine cannot; what it can do is to carry out complex mathematical analyses much faster than man, providing the material is fed to it in a digestible form. Here is one snag – how to arrange information in the right way so that it can be dealt with mechanically. Probably because this is often so difficult in medicine, computers are misused. The value of speed deludes people into acquiring a great deal of useless information, which they may not know what to do with.

However, the computer can:

- Carry out mundane and boring chores (such as statistics), but the man who puts data from a small experiment on a computer just to find the means and standard deviations and carry out a "t" test is missing half the fun in research. With a sandwich in one hand, a simple digital calculator can be used during lunch-time to discover what is wanted just as quickly and more cheaply.
- Carry out complicated correlations, to look at data from several aspects at once which a researcher often cannot. Sutton's Law states that "research productivity is related to the ability to follow up a promising lead effectively". (Sutton was an Australian bank robber who, when caught, was asked why he robbed banks. His reply, "Because the money's there", states the obvious). A computer thus has a distinct place in analysis and in the search for a promising lead.
- Search the literature more effectively than the researcher, because it can look at several subject headings (linking two or more titles in a paper) at the same time, which the simple man cannot. Hence, spending £10 for a two-year search may save perhaps 5 hours of your own time looking through the Index Medicus. But, in thumbing through that catalogue of published papers something useful may catch your eye because you will be

thinking of your subject and its ramifications all the time. This the computer cannot do.

● Establish differential diagnoses, giving probabilities, when sufficient information is fed into it, and then suggest further investigations. A simulator can thus replace teachers and patients for a student, but, what a shame!

Do remember that a computer:

1. Is a high speed moron; it will do things faster than you can, rightly or wrongly.
2. Is stupid and only understands one language (you cannot talk to a computer anyway). Instructions to the computer have to be precise, specific and often in alternative terms. Each computer knows a different language (a "programmer" will help you, but will also expect to be paid).
3. Goes sick unexpectedly when treatment can be expensive and take time.
4. Follows that old square formula:

FIFO
RIFO
FIRO
RIRO

Put facts in and facts will come out (FIFO), that's fine; put rubbish in and facts may come out (RIFO), which is not so good; fact in and rubbish out (FIRO) isn't good either and may not be realised for a long time; rubbish in and rubbish out (RIRO) is no more than you deserve.

5. Remember too that someone has to pay for sophisticated tools.

CONCENTRATION

Concentration is the key to results in research. Yet no other principle for effectiveness is violated so constantly. The common motto seems to be "let's do a little bit of everything"; it doesn't work. It isn't easy to concentrate yet mostly it is a question of simplifying and distilling the statement of objectives.

The man who is immersed in day-to-day problems has no time to plan for the future, so that it becomes mentally and physically impossible to concentrate, except for the most industrious. W. B. Cannon wrote in 1945, "Take away his free time and the researcher is utterly destroyed as a contributor to knowledge" (The Way of an Investigator. Norton, New York).

It is too easy to dissipate energy in a multitude of unfinished and unpromising projects. So the first advice is, don't try to do two things at once. Hence, if you are driving at 90 m.p.h. (which you should not be) and about to pass another car, then don't look down to see the time. Be content and wait five seconds because time does not matter for the moment. So too in experiments. Those who wander from their objective never get there; they confront unexpected obstacles (of their own choosing) encounter difficulties (of their own making) and, finally, run out of steam (and why not?); all because the objective became cloudy and less precise than when the work began. This need not be so, even though one of the hazards of research is that new ideas come during an investigation and try to modify a scheme.

Discipline, mental discipline, is the answer, but unfortunately it is unacceptable to many. A little loneliness is necessary for clear thought and may be the only way to concentrate.

CONCLUSIONS

At the end of the day you will have to collect all the facts, fit them together, sort the important from the trivial and come to a conclusion, often a painful process. It is worth remembering that observations are always correct, conclusions rarely so.

However slight the research contribution may be, everyone should state one or more conclusions. There are 3 possibilities:

- The conclusions agree with what everyone already knows and in this manner support contemporary knowledge. This is the commonest finding, but no less worthy for that.
- The conclusions sustain ideas and views suspected to be true but for which there was previously little corroborative evidence. Reputations are built this way.
- The findings are entirely new and have not been thought of before. Fortunate indeed is the investigator whose work comes into this category. The common response is disbelief, so that a further publication in defence of the new concept becomes imperative.

In reaching conclusions, the author must survey the whole of his work, its intention, execution and the results obtained, because conclusions represent a synthesis and interpretation of data, a mature and relevant assessment. They must be justified by:

1. Comparison with theoretical knowledge.
2. Comparison with results obtained elsewhere.
3. The efficiency of the apparatus and methods, and explain any reasons for discrepancies or variations.
4. Some computation of human errors, environmental and selection factors (the use of statistics when necessary).
5. Finally, relevance. Results and conclusions from work on rodents do not necessarily apply to Man.

A common conclusion from work done is often a new hypothesis, perhaps the most valuable lesson of all.

CONTROLLED TRIALS

A peculiarly British innovation for making unbiased comparisons, the technique of a controlled trial has wide applications in medicine, industry, education and administration. It is rarely used except in clinical medicine where it has proved to be a sharp and efficient tool. Indeed, it is now the only recognised method for assessing the value of a new drug or of a new treatment: so much so that few reputable journals will publish papers if the trial has not been controlled. What then is meant by a controlled trial?

Looking at old notes is no substitute for a prospective trial, but it is often the only way in which a new thought or impression can be given substance and so developed into an hypothesis for a prospective study. However, it is no good improving the quality of the data if there is inadequate self-criticism and this is the reason why all postgraduates are encouraged to participate in research. The question is, what research? The controlled clinical trial is a research tool readily available for anyone to use. It has a built-in method of criticism (by using statistics intelligently) but does demand a fairly strict discipline from the investigator.

WHY DO A TRIAL?

There are five reasons:

- Because there is genuine doubt about the efficacy of two or more different treatments. Once acceptance of a method of treatment has become established, it is usually too late to plan a trial, for it may then be unethical to do so.
- To show that a new therapeutic measure is an advance on the old, and demonstrably so.
- The furtherance of new knowledge about the detail of a disease or its behaviour in certain conditions.
- As a method of finding out more about a condition which

occurs unequally in different populations, or sexes.

- For self-discipline, so that personal prejudices and preferences are subordinated. A clinical trial is not merely an intellectual exercise, similar to the discovery of mathematical truths, for one must acquire a certain amount of knowledge first. The object is to collect facts and derive conclusions from them. No-one can tell, by the use of reason alone prior to the actual experimentation, which is the better treatment; this can only be ascertained by trial and error over a period of time or by a properly performed clinical trial.

The discipline is not just in carrying out a trial properly, but in the planning and writing or a protocol beforehand. One has to define:

1. the drugs to be used,
2. the disease to be treated, and
3. the population of patients who will enter the trial.

It may be advisable to plan and carry out a small "pilot" trial before starting the larger planned trial.
There are 5 important general principles:

- The trial should be planned to give a decisive answer to one or two questions which will be of value in the treatment of subsequent patients. The more precise the question, the more precise will be the answer.
- Patients must be selected for inclusion in the trial by specific criteria.
- Patients must be allocated at random to the "treated" or "control" group before it is known which treatment they will receive.
- Systematic observations must be made on all patients.
- The physician and patient should be unaware which treatment is being given, if at all possible: that is, the trial is conducted by the double-blind system. A "blind" trial is when the patient is unaware of which treatment is being used, and a "double-blind" is where the patient and physician are both unaware of which treatment is being used.

PLANNING

This can be divided into two parts:

- The primary objectives – that is, to answer 2–3 important questions.
- The secondary objectives – to answer some less important but perhaps interesting questions.

In acute disease one might pose three questions:

1. Is the mortality reduced?
2. Is the morbidity reduced?
3. Is the duration of the disease shortened?

In chronic diseases, the three questions could be:

1. Does the drug affect symptoms and signs?
2. Does the drug cure the disease?
3. Is the patient improved by the drug?

SELECTION OF PATIENTS

The criteria for entry of a patient into a trial must be clearly stated. These usually specify a defined age group, the presence of certain signs and symptoms, the duration of the disease and which patients are to be excluded.

Once a patient is selected for entry into the trial he is allocated to the treatment group or the control group at random. The random allocation should be provided before the trial begins by using a series of random numbers to be found in any book on statistics. For instance, in testing a new drug, a special prescription form should be completed for each patient, so that the pharmacist (who holds the key to the identity of the drug) can dispense tablets or ampoules acording to the sheet of random allocations. The sheet will look like the table on p. 40 (where T = treatment by the new drug, and C = control group).

It will be seen from the master sheet that the next patient to enter the trial will receive the new drug – but this information must not be available to the physician or to the patient. It should also be noted that the allocation has been restricted to groups of 10 so that there are equal numbers of "treated" and "control" patients in every 10 patients seen. In the next 10 patients, numbered 11–20, the allocation will be different and the sequence might be C, C, T, C, T, C, T, C, T, T. All these measures are taken to exclude bias on the part of the observer.

Serial Number	*Patient's Name*	*Age*	*Sex*	*Treatment*	*Date*
1	A. Smith	56	M	T	9.8.74
2		48	F	C	9.8.74
3	D. Jones	52	M	C	10.8.74
4	C. Brown	50	M	T	10.8.74
5	J. Black			TL	
6				T	
7				T	
8				C	
9				C	
10				C	

THE CONTROL GROUP

Some patients must necessarily be allocated to the control group. They are essential for comparison so that any doubts about the new treatment can be resolved. They receive the best treatment previously available, and are allocated at random. They should be treated at the same time and by the same physicians as the test group, and the same care in recording observations made. If the trial is "double-blind" there is no difficulty, for all patients of necessity must be regarded the same.

THE PROTOCOL

A detailed written statement of intent should be circulated to all persons likely to be involved before the trial begins and approval obtained from the Ethics Committee of your institute. Everyone must be clear about:

- What patients? Age? Sex?
- What diagnostic features must be present?
- How many patients are to be studied?
- How will information be recorded and by whom?
- Details of drug dosage, prescription, side effects?
- The experimental design of the trial.
- A flow chart showing the stages of development of the trial is valuable.

PRESENTATION OF RESULTS

When the trial is completed, the analysis of the results should be easy if the principles already described have been followed. A few tables and diagrams are usually required and the text should be in simple language so that all may understand. Simple statistics should always be done and the confidence limits included in the text.

The object of reporting the successful result of a therapeutic trial with a new drug is to influence treatment carried out by other doctors, and to demonstrate promising developments for the pharmaceutical industry. The written record should thus be clear, concise, crisp and accurate. The conclusions must be justified, understandable by all and in language which can be accurately translated into foreign tongues. Medicine is international and advances in therapy important.

THE GOLDEN RULES

For the completion of a successful clinical trial these are:

- The observer should be in ignorance of what treatment is being given.
- No enthusiasm, but some emotional involvement is necessary for the drive to finish the work. The object is to discover the truth, not to prove our own ideas right.
- Make only one change at a time.
- Assessments are made by the same person.
- Allow time for patients and doctors to settle down.
- Decide the important questions to be answered before the trial.
- Measure accurately.
- Assess symptoms and signs separately.
- Use the correct dose of the drug.
- Always allocate at random.

What we require are FACTS to give a decisive answer one way or the other (a mnemonic emphasises this):

Formulation of the problem
Allocation without bias
Comparable groups

Therapeutic evaluation
Significance of what has been found.

When a man thinks for himself, he has become a philosopher. Tradition, custom and habit will not deal with new therapeutic measures – be they new drugs, new surgical operations, new methods of management – which have to be put to the test. The difficulty comes when we ask, what test? The controlled clinical trial is one method of testing, providing one recognises four things:

- that bias is an inherent human weakness;
- that the person who believes sincerely that what he does is correct, cannot have doubts and therefore cannot undertake such trials;
- that the necessary statistics must be used in the correct context; they are usually quite meaningless unless randomisation has been the method of allocation;
- that if it is ethical to start a trial, it is usually correct to finish it.

There are several good books and papers to consult. The British Journal of Anaesthesia (1967, *39*, 277–319) devoted a whole issue of various papers on the subject. Two books on Controlled Clinical Trials (Witts 1964; Flamant, 1970) should be read to appreciate the breadth of the subject, and a paper by Dundee (1974, *Proc.Roy.Soc.Med.*, *67*, 586–588) describes the practical aspects of the changing protocol before the real trial gets underway. Consult also "The design of forms" (1972. H.M.S.O.) at the planning stage.

CONTROLS

Much research depends on comparisons. Reichmann (1961) says that "measurement is by definition impossible without comparison". The difficulty comes when we ask, "Comparison with what?". The importance of controls has been dealt with under "Controlled clinical trials", but here I want to emphasise more strongly the nature of comparative controls in every form of research activity because what constitutes controls is poorly understood by many workers.

If a new therapeutic agent allows people to survive from an otherwise uniformly fatal disease (Streptomycin for miliary tuberculosis, penicillin for streptococcal septicaemia), then controls are probably not required and once the agent has been shown to have a life-saving effect it would be unethical to plan future experiments in that way.

However, such good fortune for the investigator is rare; he has to have controls by which to compare a new treatment. In studying sick persons a healthy young adult is not a control, nor may the results of laboratory animal studies be so considered. For surgical procedures the provision of controls may be much more difficult, particularly if there is no alternative manipulative procedure available (a "sham" operation would be unethical anyway).

In reading publications introducing new concepts and new therapies, one should examine the nature of the control patients with special care, because –

- controls should be managed at the same time as the treated group (concurrently)

Patients seen earlier or later are not controls in the honest meaning, because with time people and things change (remember leeches and blood letting for hypertension? It was not so very long ago).

- controls should, indeed must, receive the best available treatment available when assessing the value of a new therapy.

Often it is quite unethical to compare patients, say, receiving aspirin for rheumatoid arthritis with those having a placebo tablet. That is not the way to treat a patient with rheumatoid arthritis!

- the person making the measurements should be unaware of the treatment given, if only to exclude bias but also to make sure that all recordings carry the same weight of accuracy, interest, diligence. Clearly, a double-blind scheme, where patient and investigator have no idea which treatment has been given, is the best; unfortunately, it is not always possible to provide and therefore the results must remain questionable.

In any case, one of the best checks of the reliability of an experiment is to show agreement with previously established results in that field. Hence it is often worth including a treatment solely with this object. When the results of the standard treatment in the controls differ significantly from that generally accepted there is likely to be some hidden bias in that group.

CONVERSATION

Conversation during meals and at other times is one form of communication which should not be avoided. There are five kinds, in diminishing order of importance:

- The opportunity to discuss the appraisal of new ideas is essential if only because, as Somerset Maugham once wrote, "It is a terrible thing to have no means of expressing oneself, always to have to keep one's feelings a secret."
- Exploration and evaluation (two rather unctious words) of new people, their personalities, ability, habits – essential if you are to know who will help in an emergency, who can be telephoned at short notice and will act.
- The exchange of interesting and unfamiliar general information is not just important, it is absolutely vital for sparking-off latent ideas in your own mind.
- The routine delivery of necessary information for the successful co-operation in some joint exercise, to keep others informed of what is going on.
- Discussion of the daily round, sport, the weather, T.V. and so on, may be passing time rather than using it, but is no less important for that. It is recognition that tittle-tattle is part of human relationships, an essential facet of good morale within an institution if not within the research team. Personal dedication is acquired by such simple means.

COSTS

Every research project will have to be costed, preferably before work starts otherwise deep water is ahead. If application is made for funds then costing has to be detailed and realistic: most people do it badly. In the long term you cannot afford errors in money matters.

In industrial practice, costs are broken down into 4 main categories and these will serve our purpose well:

- Production costs, that is the money required to provide value to the customer. These include all the expenses on staff, laboratory equipment and essential necessities to complete the research project. They can usually be estimated fairly accurately because the salaries of staff are known and so is the price of equipment. What is not known at the beginning is how long the research will take and what new staff may be required in, say, a year's time.
- Support costs provide no value of themselves but include the cost of administering a grant (5% of the total is the estimate at the Royal Postgraduate Medical School), travelling to talk to others in the field, making slides for lectures, reprints of publications, a secretary's time in producing a report. Most grant-giving bodies expect these costs to be absorbed by the institution where you work and prefer to look the other way when the subject is mentioned. Even so, they may represent an appreciable sum.
- Insurance against doing the wrong thing. Using a library, photocopying important publications, the purchase of books, even rechecking data and computer time.
- Waste costs, or efforts which do not produce results. The most expensive waste is usually not doing something within the time available, and the best way to find waste is to deliberately look for it.

Most grant-giving bodies recognise the costs of 1 and 4 but

refuse to allow for 2 and 3. The only recourse is to apply for a moderate sum to cover the unexpected, set up a contingency fund or make further application when the work is underway and has become productive.

COURAGE

It takes courage to stand up and speak and another sort to sit and listen. If research means anything to you, you will learn to do both and accept the consequences.

CREATIVITY

Exciting research depends largely on creative thinking. It is not easy. It may take a year to think correctly and constructively and only a few months to carry out the work to support a new idea. It is important to find time to look out of the window and dream, time rarely wasted for an active mind. Time has to be made, fought for and used to best advantage, for no-one will give it to you freely and dreamlike activities will be resented even by those who should know better.

Colin Blakemore, writing in "Illusion and Art" (1973 Edited by R. L. Gregory and E. H. Gombrich, Duckword, London) said that, "The brain, a strange grey substance that fills the skull, is a biological instrument more complex, more compact, more sophisticated than any machine made by man. Within its enigmatic kilogram-and-a-half of jelly resides a power of computation that embarrasses the weightiest computer, a power of communication that out-performs all telephone systems in the world and a power of creation that has generated the Mona Lisa on the one hand and nuclear weapons on the other". Surely a very complete description of what creativity is all about.

There are 10 guide lines:

- Define the problem as accurately as you can. As Bronowski once wrote – "Ask the right question and the answer is often obvious".
- Find out the relevant facts, often difficult, and this job alone may take an inordinate amount of time.
- Analyse the problem and what you want to know about it; write it down, look at it often.
- Use your imagination by asking What is different about this? How can it be done? What is the purpose? What are the applications?
- The association of ideas: where else would this apply?
- Sleep and think.

- Then criticise the idea. What are the weaknesses? What are the difficulties in starting to demonstrate the truth?
- Read in earnest – anything and everything, good and bad, on the subject.
- Listen to lectures on related subjects. Don't look for standard answers but question them until you are satisfied; if you are not, then add further to your list of questions.
- Learn to think creatively by questioning (and answering) every piece of doctrinaire teaching. It is a painful process, but once acquired you become someone of importance.

Action is more noticeable than thought, hence the "doers" are more impressive. However, there is no action without thought: unfortunately creative people are often inactive physically and so considered lazy. Do not be caught out just because the creative person has no competitive spirit, it does not mean that there is no contribution. The people at the heart of research, the creative people, are indispensible. Whether they know what they're doing is another matter, for there is no yardstick by which to judge creative work. The director of the research team, therefore, has to decide.

One person will regard a problem in research as an opportunity and will seek to exploit it; another regards it as an enigma and will try to explain it. So when the creative man and the ordinary researcher talk of challenge they refer to different things, but the root cause is the same. Nobody knows how research works. I don't.

Creativity in any aspect of life must have relevance if it is to be valid; in research this is usually the relevance of application to current problems. Research means action, that is, doing something; rarely is it to do with thought alone. Thought is required, but relevant action must follow. Not everyone sees it that way so the problem is how to keep creative people working. There are 10 ways to help:

- Make sure that their jobs are secure and make them understand so.
- Give them freedom from emotional disturbances – and this may mean taking them away from clinical work for a time.
- Avoid loading them with routine administrative problems which can easily be dealt with by others.

- Recognise that creation requires a great deal of mental effort, which may be greater than purely physical effort.
- Creative people should not be isolated, so ensure that they are still in the main stream of work and not up a creek.
- Insist on reasonable hours of attendance: the great danger is that someone comes in at midday but works until midnight.
- Don't be too soft or too rough on them, bring them down to earth and insist on regular reporting.
- Allow time for them to mix with their own kind, by attending scientific meetings, visiting other units, travel.
- Try to involve harmony of the practical and the perishable by consultation and application of new ideas. It's not easy, but who wants an easy life? Three great creative ideas of science have been:

1. the concept of order in an apparently disorganised world;
2. cause and effect: that every effect has a cause;
3. that probability and not certainty is a measure of events.

All are applicable to research projects whatever their nature.

- Finally, tact. If you discover how to do away with 50% of the nurses in a hospital, keep quiet. If you announce what you are going to do, the opposition will have time to construct real and imaginary reasons why you cannot.

CRITERIA OF A RESEARCH PROBLEM

Many people wish to do research but fail to realise that the problem under consideration must be formulated so that it is possible to use the scientific method and obtain an answer.

- Ideally it should concern a natural phenomenon which is quantitative and preferably amenable to experimentation.
- The subject should be capable of being stated as several alternative hypotheses. It should therefore test some proposition in which there is a real difference of opinion.
- It may deal with some little known relationship or some neglected question. Significantly, it must arouse the curiosity and interest of the investigator.
- The primary object is to obtain new facts in an important field, not to repeat what has been done before. Research should have some clinical application, either directly or indirectly; in the past, much purely laboratory investigation has failed for want of clinical direction and there is still much pointless research undertaken, however well dressed up it appears.
- The research project must be circumscribed, specific and quite definite in its objectives. Beveridge's (1951) *The art of scientific investigation* and Wilson's (1952) *An introduction to scientific research*, are recommended reading at this stage. Claude Bernard's (1878) *An introduction to the study of experimental medicine* is also obligatory reading.

  The wise man will start a card index of possible subjects for research, and carry a small note-book in his pocket in which to record ideas as they occur.
- Finally, the research project should be adapted to the material available and be within one's capability. It is useless for anyone to plan a project which, for instance, requires a great deal of histochemistry if he knows little of the subject, and has no expert help. Learning the techniques alone may take 6–12 months, during which time the planned research has not

advanced at all. Much research reviews old problems using new techniques, so that learning new techniques may be the way to open up a profitable field of discovery; but time must be allowed to acquire the necessary skills before starting a project.

CRITICISM AND SELF-CRITICISM

It is no good improving the quality of data if there is inadequate self-criticism. This is the reason why medical schools introduce students to the problems of research and statistical evaluation and why all post-graduates are encouraged to participate in research. No-one likes criticism, for as Somerset Maugham noted: "People ask you for criticism, but they only want praise".

Much research is reported by enthusiasts and therefore by people whose critical ability is impaired. This is clearly demonstrated by later publications on the same subject when the original findings are not supported or a different interpretation put on them by others. Moreover, even for those who write repeatedly on the same topic, the author begins to hear his own echo and loses some of his enthusiasm. Woglam put it neatly when he said: "The greater the experience of the investigator, the less successful the results are apt to be".

Self-criticism usually means an ability to reject any data that do not satisfy the strict criteria of a professional researcher and to be prepared to say "No" when the evidence does not justify the conclusions made. It requires courage not conviction, but is the spur to acquiring new information. For some research projects, often those which have been topics of promise for half a century, it may not be possible to obtain more information simply because the techniques for collecting it are not available. Either you can think of an entirely new approach quickly or you are wasting time which could usefully be employed in another field.

When we talk about criticism we refer to work done and not to personalities. Ashe (1972) finds three kinds of critics of written prose; the quick-witted, lively and questioning type; the slow and uncertain; and finally the person with an eye for the comic, incongruous or unfortunate phrase, the man who will laugh when he's not meant to. All three are to be found in research. So let's examine a published paper to see how criticism works.

Having analysed the work of others the lessons can be applied even more strictly to one's own efforts. There are 5 areas to examine, with a high index of suspicion:

- METHODS

Were the methods used acceptable and reliable?
What kind of inherent errors do they have?
Are the methods now out of date?
Have you better ones?

- MATERIAL

Was the sample representative of the population or just a selected group of hospital patients?
What sort of controls were used?
Was the material (patients, animals, specimens) allocated at random (and therefore a probability sample)?
What comparisons were made? Were they justified?
Has there been selection?

- TABLES

An accountant looks at the profit and loss of a company to find its true status, ignoring the written account by the chairman of how well things have gone in the past year. In the same way a researcher will want to know how to read a table which can be consumed in considerably less time than it takes to read about it. Read the title carefully; a good title tells precisely what the table contains. Read any other explanation about the source of the data and any footnotes. Then:
1. examine what units were used
2. look at the overall average
3. see how much variability there is (the confidence limits)
4. now look for things you weren't looking for (aberrations, anomalies, irregularities) because these are often worth pursuing; one of the secrets is serendipity, from which fruitful findings in research so often come,
5. finally, look at the statistics used: were they applied correctly?

- GRAPHS, HISTOGRAMS, CHARTS

Do these provide extra information? Are the curves of a chart justified by the number of observations or has there been some artistic licence in their construction? Are they capable of a different interpretation from that provided by the writer?

Are the axes of graphs and histograms labelled correctly and the intervals comparable? More importantly, does the plot of a graph pass through zero point, and if not why not?

● CONCLUSIONS

Are they justified by the evidence provided?
How reliable are they?
How important are they?

Having learnt how to criticise the publications of others, now is the time to apply these probing questions to your own work. If you do not, someone else will.

DECISIONS AND DECISION MAKING

Many people in research shy away from making decisions, or delay until they are ineffective. Research is an executive pastime. Decisions have to be made; sometimes they are right, but then no-one is infallible. One plays the game to win. Creative man lies in a world of uncertainty, with imprecise and constantly changing features. Yet some of these can be changed and that's what research is all about. It is important, therefore, to think in terms of probabilities rather than certainties (see Statistics).

It is often suggested that the executive man (a surgeon?) is aggressive and extroverted, whereas the creative man is shy and introverted, because he has to live within his own head! This contrast between "thinking" and "doing" is over-exaggerated. It has also been said that the executive lives in the present, the creative in the future. This is surely untrue. Both are practical people, but one has a bit more imagination because many inventions which at first seem impractical, turn out to be eminently useful.

We all have to make decisions daily. Some are simple and straightforward, such as when to cross the road. We look left and right to see what traffic is coming (observation), calculate the distance between vehicles and us, work out their speed of approach compared with our own ability to cross from pavement to pavement, then come to a single conclusion "cross now, or wait".

Many decisions in research are of this order, but some are more complicated. Should the research continue or finish? Can further funds be applied for? Should the direction of the research be realigned? These are all serious questions which require a completely different approach. In almost all cases there is more than one good solution to any such problem, so write down in some detail the reasons for making your decision. Such notes are invaluable later in life when the same problem occurs again (as it will) under different circumstances and where earlier errors can be seen in a better light. Most decisions are made on the basis of self-advancement, but an adequate knowledge of people, concepts and the things involved in making a good or bad decision, are terribly important. It is essential to have a wide knowledge of the effects of your decision on the group with which you work and the society in which you live. Often the facts seem to point to a clear choice but outside considerations may limit the alternatives or force you to accept a different decision. What guide-lines, then, may help you to make up your mind? The scientific method can be adapted to this problem too, for it has become recognised even in industry as the best method:

- Define the problem in precise and accurate terms; difficult often, but essential if a good solution is to be found. People often talk of the "right solution" but this is rarely true in complex situations; a better term is the "best solution at the time".
- Collect as much data as is reasonable about the problem.
- Weigh the data for its value and relevance to the problem under consideration.
- Frame alternative solutions based on the data.
- Select the solution which appears best from the available data.
- Decide how to act on the decision (and consult "COMMITTEES").

If your decisions affect the group with whom you work then full discussion must be allowed so that it becomes clear:

1. Who is willing to stand up and be counted, that is, who is really interested in finding a solution;
2. Who is willing to listen to other points of view, not necessarily your own;
3. Who is anxious to explore real differences or areas of disagreement and who is ready to account for the results;
4. Finally, the effectiveness of leadership in setting the climate, so that people are free to participate, is vital. The leader can state the problem, establish the rules of procedure, set the limits of delegation, state the problem in different terms and pass judgement on the data, while chosing an acceptable solution

and providing for future action. Even so, decision-making is an art, not a science, and like all forms of art can be learnt by experience and diligent application (but for the lucky one there is something innate).

DELEGATION

Delegation means investing another person with authority to act on your behalf, with the minimum of restrictions. It does not mean giving others work to do while retaining authority and accountability, nor does it mean asking others to do the work in the way you would prefer; both of these are unduly restrictive. By delegation a chief extends his powers and the chance of success and, more importantly, subordinates learn to grow by experience and responsibility. Passing on responsibilities does not absolve the director of a team from accountability – it merely spreads the load; all are involved.

It is also necessary to pass on the right to make decisions, take action and give orders. Hence, clear definition of the area of this delegated authority is important, and should be in writing so that a subordinate knows exactly where he stands. Conditional authority is not true delegation. On the other hand, a person should be accountable only to the person who delegated the authority and responsibility and to none other. In addition, some form of control will be required and the amount of delegation should be related to this.

DIAGRAMS

Some people think with a pencil in hand. Their ideas are immediately translated into a visual image. At the Royal Postgraduate Medical School we have a large workshop full of engineers. Go down there with a simple sketch of a piece of apparatus, explain what has to be made and the Chief will turn this into an engineer's drawing for approval; it may then be difficult to follow because the expert has altered the emphasis, but the piece of equipment will be made for use. It may have to be modified, again another diagram to explain why.

Imagery, thus, can be expressed in language and drawings, preferably both. Hence the value of simple diagrams in all research, to reinforce ideas. The professional artist can illustrate things that photography cannot, either of the real world by three-dimensional effects or of abstract ideas. Indeed, the preliminary examination of most data is facilitated by the use of diagrams; they prove nothing, but do bring outstanding features to the eye. They are valuable in suggesting critical tests to be applied to the collected data, perhaps explaining the conclusions found on such data and may find an important place in the writing of a worthwhile publication. Not only may diagrams act as a personal aid, as a visual stimulus to writing clear prose, they may also:

- aid intelligibility of the publication, because diagrams are particularly useful for illustrating apparatus and directions of flow in fluid systems.
- provide an economy of words in the script (see the book by Lockwood).
- present definitive ideas in an artistic medium (minimal lines may stimulate the imagination of the reader).
- crystallise one concept more clearly than words, but the diagram must be simplified until its message becomes instantly understood.

- provide the researcher with ideas for the future. A notebook of diagrams (with the dates) may illustrate in retrospect better than words how and where the investigation has gone wrong.

DICTIONARIES

Most people use a dictionary as they would a telephone directory, to look for a specific item. Why not browse through a dictionary (as a novelist does the telephone directory to find the most appropriate names for his characters) just to find the unusual, the descriptive and the pointed word? The value of Roget's Thesaurus is not so much that it provides a choice of alternative words for the same thing (it doesn't, of course, because different words have different connotations in English) but that it makes you think. Suddenly, the right word explodes out of your head.

Hence, keep both beside you as you write, as good companions in the labour of composing a gem for publication. The finicky (and the aged, for they were brought up on it) will ask you about Fowler. Yes, that too, but it does require time to delve and perhaps it is the book to consult when the writing is done. It is good bed time reading rather than desk consultation stuff.

The intelligent writer will have a good dictionary on his desk for correct spellings and for the recognised definitions of words in his native language. The *Concise Oxford Dictionary* is small and manageable, but for more precise definitions and wider examples, the larger *Shorter Oxford English Dictionary* in two volumes is recommended.

When it comes to definitions of medical terms and abbreviations, the following should be consulted:

The British Medical Dictionary (1961) Edited by A. S. MacNalty (Caxton, London).

Butterworths Medical Dictionary (1965) Edited by A. S. MacNalty (Butterworths, London).

Dorlands Illustrated Medical Dictionary (1965) 24th Edn. (Saunders, London).

Stedmans Medical Dictionary (1972) 22nd Edn. (Williams and Wilkins, Baltimore).

Dictionary of abbreviations in Medicine and related sciences (1971) Edited by E. B. Stein. 3rd Edn. (Balliere, Tindal and Cassell, London).

A useful small volume to examine is by F. Roberts, *Medical terms, their origin and construction.* (5th Edn. 1971, Heinemann, London) and, another, the *Alchemist Glossary of Medical Terms* (2nd Edn. 1963, Alchemist, London).

DIRECTOR OF RESEARCH

Every research team will need a director, the man who finds the money, plans projects, writes reports, is responsible to Higher Authority. The good Director of research, that is the man in overall charge of a particular project or team, has special qualities, for he:

- has the power to use people to the best of their abilities, for their own satisfaction;
- has the ability and imagination to organise the work;
- knows how to work with others. People do not have to like each other to work together, but they must gain respect for each other.
- must be the research team's first man (and second and third);
- must be able to observe rules and have the judgement to know when to break them;
- recognises that good direction is tenuous. If he's sure his solution is the correct one he must hold on to it, come what may.
- knows how to create powerful images of the importance of the work in hand;
- should have a sense of design of the project;
- must be as competent with his hands as the best of the team;
- has a watch and a calandar and studies both;
- knows how to finish strongly;
- should be worried about grant monies all the time. He has to create results with an eye on the clock, to realise where the money is going; the idea of "cost-control" is new to most in medical research.

Perhaps these 12 features are those of the ideal organiser. Maybe, but we all have to try. Worse still, the director has to discern the difference between quality and quantity, where often the latter is the factor which makes the demand for further money a little easier. He also has to be a "Futurologist" in the sense that he has to look at the past and the present and then decide what the future will be. Too often, he will assume that changes will occur in the same direction, being aware that no one can stand still in research but has to move forward; the right direction is the important thing. He will not necessarily be the leader, but it helps if he is that too.

DISADVANTAGES OF RESEARCH

There are many. For a start, research takes time, often an exorbitant amount of time, in thinking, planning and execution. As a result, something has to go – social events or work – simply because no one has managed to invent the 26 hour day. Whichever is discarded it should be acknowledged by the researcher and those around him; to run with the hounds and the hare has never really worked in practice, at least not for long. It's a tough decision and putting it off does not make it any easier.

Research often sounds romantic; it sometimes is but in between there is a four-letter word, WORK. Hard work is required for success in any project; some people simply are unable to put in enough to draw out any treasure. A great deal of research is repetitive, mundane and sometimes boring. There are periods of acute depression when things go badly and others of great excitement when things go well. The big "Breakthrough", found in films, T.V. and newspapers, is a rare event.

Worse still, there is no social security in research, no seniority, no old-age pension. Research is a harsh employer who will dismiss even the most faithful servant. The sudden loss of a research team might upset its employees, but would barely cause a ripple in the field. This is hard to accept and even more difficult to understand. No-one is blessed with omniscience and so, unless thought for the future is systematically sharpened by exploring new avenues, the obvious may be overlooked and disaster follows. There is no security of tenure in research and the risk of failure is always present. So why do it?

DISAPPOINTMENTS

Research is full of disappointments. Something always goes wrong, equipment fails to record accurately, the results don't make sense. These are all perfectly natural events in the search for new knowledge. Nature does not reveal her secrets easily and rarely to those who have never known success and disappointment almost in the same breath.

Disappointment and the ability to overcome frustration is usually called "experience". Rudyard Kipling in his poem, "If" said it all.

The complete piece is reproduced here from Rewards and Fairies, by permission of Mrs. George Bambridge and Macmillan of London and Basingstoke.

IF

IF YOU can keep your head when all about you
Are losing theirs and blaming it on you,
If you can trust yourself when all men doubt you,
But make allowance for their doubting too;
If you can wait and not be tired by waiting,
Or being lied about, don't deal in lies,
Or being hated, don't give way to hating,
And yet don't look too good, nor talk too wisely:

If you can dream – and not make dreams your master;
If you can think – and not make thoughts your aim;
If you can meet with Triumph and Disaster
And treat those two imposters just the same;
If you can bear to hear the truth you've spoken
Twisted by knaves to make a trap for fools,
Or watch the things you gave your life to, broken,
And stoop and build 'em up with worn-out tools:

If you can make one heap of all your winnings
And risk it on one turn of pitch and toss,
And lose, and start again at your beginnings
And never breathe a word about your loss;
If you can force your heart and nerve and sinew
To serve your turn long after they are gone,
And so hold on when there is nothing in you
Except the Will which says to them: "Hold on!"

If you can talk with crowds and keep your virtue,
Or walk with Kings – nor lose the common touch,
If neither foes nor loving friends can hurt you,
If all men count with you but none too much;
If you can fill the unforgiving minute
With sixty seconds' worth of distance run,
Yours is the Earth and everything that's in it,
And – which is more – you'll be a Man, my son!

DISCIPLINE

Handling people is not all that difficult. All you need is inexhaustible patience, unfailing insight, unshakeable nervous stability, an unbreakable will, decisive judgement, irrepressible spirits, a sturdy physique, plus unfeigned affection for all people – and an awful lot of experience.

There are some snags, of course, to be overcome. To control people, first learn to control yourself. Then apply the 10 principles (see PEOPLE).

ENVIRONMENT

Good research grows in a good environment. How can one provide the right conditions? Successful research does not just occur, it has to be worked for. Here are 10 ways to improve the environment:

- Group meetings to discuss progress, openly and freely: a survey of the subject or a summary of work in progress.
- Stimulating back-chat at all times.
- Encouragement from the more senior, in working, reporting, writing.
- Fostering a sense of enquiry in everyday affairs.
- A label on the door *Research Laboratory* with the names of the occupants too.
- Reading papers to scientific societies and lecturing to others.
- Inter-unit and inter-departmental meetings help to fertilise ideas and encourage all to think more deeply.
- Opportunities to travel abroad, to spread the good news.
- Interest of senior members of the same institute always helps. In the same way taking visitors around, introducing each researcher by name and letting him explain what he is doing, is a great stimulus.
- The annual social party and an annual report giving details of achievements. It may even be worthwhile to have an "open-day" when others can see work on display.

It is no accident that there are centres of excellence in institutions where the variety of research projects in different departments is all of a high calibre. Perhaps the world authority on, say, enzyme assays works next door but his very presence (or sometimes absence) helps to create the atmosphere for others to do well. In medicine, such centres are frequently those providing the best care for patients, demonstrating the value of research in producing better clinicians.

ENVY

Everyone who is successful in research becomes the object of envy and jealousy. There are no exceptions. You may not be aware of them at first, but when you are they are annoying, unbelievable, irritating and dispiriting, often unprovoked and unjustified. Worse still, the pin-pricks may belittle your success, cast doubt upon your clinical ability, imply that the work is not important, question your real understanding. Frequently they come from colleagues and people whom you thought friends; a sad reflection on human nature.

The dictionary makes a distinction between the two words, defining envy as "a grudging contemplation of more fortunate persons", and jealousy as "resentment towards another of known or expected rivalry" but to the recipient the difference is of little matter. What can one do? First of all, understanding. Jealousy and envy are normal human traits so make allowances for them, particularly if the information comes to you second-hand. It is important not to brood over snide remarks nor let them upset you, for the result is inevitably a lower standard of work and a smaller output, both signs of defeat. There are, however, three positive things you can do:

- Be humble in putting forward new ideas, by giving credit to those who set you on the right path. By "spreading the load", as it were, it becomes more difficult for your detractors to appear other than they really are.
- By appearing completely unaware that anything so unprofessional as jealousy should be possible in the subject in which you work for the common good of all. The Christian attitude, to turn the other cheek, is right but difficult in practice. Always greet the person whom you know to be most disturbed even though there may be no reply. To forget is to forgive.
- Make them really mad by turning out even better work next time, so that with a world-wide reputation the jealous one

begins to look silly and petty (which he is anyway). All this means a great deal of extra labour with painstaking accuracy, diligent application, well written publications, clear lectures and often a great deal of courage when things look very black. Never give in to moral blackmail. No good, either, will come from bitter arguments; far better to let the facts speak for themselves as they usually do sooner than expected.

ETHICS

This is a code or set of principles for dealing with patients, essentially the difference between what is good or bad for the individual patient. Some of the moral conundrums, with which ethics deals, are not analogous to scientific questions. In a scientific issue, when all the relevant facts have been gathered, we can in principle always decide correctly. This is not so in a moral situation.

For guidance the Helsinki Declaration was issued in 1964 and is reported on the next page; since then the Medical Research Council (1972) the Royal College of Physicians (1973) and the Canadian Journal of Nursing (1973) have also pronounced. There are many ethics. J. S. Mill proposed Utalitarianism, that is, the greatest good for the greatest number, which may apply to medicine in general; but most doctors are Hedonists at heart, arguing for the greatest good for the individual. Even so, the protocol for every proposed clinical trial and every human experiment should be submitted to the Committee for ethics at your institution before any treatment is undertaken. In the old days it was sufficient to ask, "Would I allow this treatment on my own children?" Today this is not enough because in the search for truth, some investigators fail to distinguish between what is good for the patient and what is good for publication. The thirst for knowledge sometimes has to be curbed.

Higgins (*Arch. Surg.* (1971) 102. 160) goes further and makes five good points, that it is unethical:

1. to attempt to analyse the effects of any treatment given in an uncontrolled and haphazard manner;
2. not to carry out a properly controlled clinical trial when there is doubt about the value of a new treatment over that in current practice;
3. to conduct a trial without the appropriate clinical and statistical design;

4. to carry through a trial without proper patient care, understanding and permission from patients and colleagues (see ETHICAL COMMITTEE);
5. to terminate a trial unless a statistically significant result has been obtained.

DECLARATION OF HELSINKI, 1964
It is the essential mission of a doctor to safeguard the health of the people and his knowledge and conscience are dedicated to such fulfilment. The Declaration of Geneva of the World Medical Association reinforces this with the words "the health of my patient will be my first consideration" and the International Code of Medical Ethics declares that "Any act or advice which could weaken physical or mental resistance of a human being may be used only in his interest". Both follow the Hippocratic code. Because it is essential that the results of laboratory experiments are applied to further scientific knowledge and to help suffering humanity, the World Medical Association prepared the recommendations as a guide to each doctor in clinical research; the standards as drafted are only a guide to physicians all over the world and do not relieve doctors from criminal, civil and ethical responsibilities under the laws of their own countries.

In clinical research a clear distinction must be recognised between two aspects, where the aim is essentially therapeutic for a patient, and that which is purely scientific, but without therapeutic value to the person subjected to the research. The declaration is in three parts and I have added comments in brackets.

A. *Basic Principles*
1. Clinical research must conform to the moral and scientific principles that justify medical research and should be based on laboratory and animal experiments or other scientifically established facts.
2. Clinical research should be conducted only by scientifically qualified persons and under the supervision of a qualified medical man.
3. Clinical research cannot legitimately be carried out unless the importance of the objective is in proportion to the inherent risk to the subject.

4. Every clinical research project should be preceded by a careful assessment of the inherent risks compared with the forseeable benefits to the subject or to others.
5. Special caution should be exercised by the doctor in performing clinical research in which the personality of the subject could be altered by drugs or the experimental procedure.

B. *Clinical Research Combined with Professional Care*
1. In the treatment of a sick person the doctor must be free to use a new therapeutic measure, if, in his judgement, it offers hope of saving life, re-establishing health or alleviating suffering.

Whenever possible the doctor should obtain the patient's consent, freely given, after a full explanation of what is proposed. In cases of legal incapacity the permission of the legal guardian replaces that of the patient.
2. The doctor can combine clinical research with professional care, for the acquisition of new medical knowledge, only to the extent that clinical research is justified by its therapeutic value for the patient.

C. *Non-therapeutic Clinical Research*
1. In the purely scientific application of clinical research carried out on a human being it is the duty of the doctor to remain the protector of the life and health of that person.
2. The nature, the purpose and any risks of clinical research must be explained to the subject by the doctor.
3a. Clinical research on a human being cannot be undertaken without his free consent, after he has been fully informed; if he is legally incompetent the consent of the legal guardian should be procured (but this is to be deprecated).
3b. The subject of clinical research should be in such a mental, physical, and legal state so that he/she can exercise fully the power of choice.
3c. Consent should, as a rule, be obtained in writing. However, the responsibility for clinical research always remains with the research worker; it never falls on the subject, even after consent is obtained.
4a. The investigator must respect the right of each individual to safeguard his personal integrity, especially if the subject is in a

dependent relationship to the investigator (and this applies to members of a research team).

4b. At any time during the course of clinical research the subject or his guardian should be free to withdraw permission for the research to continue. The investigator or the investigating team must discontinue the research if in their judgement it may be harmful to the individual.

ETHICS COMMITTEE

Every experiment proposed for Man should be submitted initially to the Ethical Committee of the institution where the work will be done, a procedure as important for the reputation of the institution as it is for the experimenter.

Years ago this was not necessary, not because people were unadventurous but because things have changed. We now have new, sophisticated and sometimes dangerous methods of investigation, powerful drugs, worried doctors (see H. M. Pappworth's *Human Guineapigs*, Routledge & Kegan Paul, London) and an enlightened public, patients protection societies and things like radioactive isotopes, which may contaminate the environment. All these demand that every procedure should conform to a certain code of conduct; what that conduct should be will be defined, although not in precise terms, by the Ethical Committee. Who should serve on this body? The answer is not as easy as one might think for members will have to serve the needs of society, of advancing knowledge and of the individual patient, yet keep one eye on the legal aspect too. The committee should be a decision making body, accountable to higher authority, yet avoiding impedance to the march of progress; usually it will contain:

- The head of the institution because he is ultimately responsible for its good name and public image.
- A representative (two or more) of those actively engaged in clinical research, if only to vet the design and competence of the experiment.
- A representative of the nursing staff, in a hospital, so that those who look after patients from day-to-day may have some voice in how things are conducted and be in a position to make important observations on the effects of experiments.
- Someone to represent the general public. Here is one difficulty because the layman may be quite capable of making a

favourable decision (for example, who is to receive a precious kidney transplant) but emotionally unable to make an unfavourable one (to decide who will be the unfortunate person not to receive a kidney and therefore must die). For most procedures he must be aware of the risks involved either way (whether the patient has a drug or not) and so must have some scientific knowledge.

- A representative of the administration of the institution so that implications for resources such as funding, space and the work-load of other units may be appreciated.
- A secretary to record and issue minutes of the applications submitted. The Ethical Committee should be the counterpart of the Home Office licensing system for animal experiments. Such records may become important if any complaint is made and proceeds to legal action. Moreover, grant giving bodies now ask for such information before allocating monies for research projects, and quite rightly too. All of which implies that a printed form should be completed by every applicant. The information should be typed so that it can be photocopied and circulated to all members before a meeting of the committee (the committee should meet frequently so that business is conducted expeditiously).

At the Royal Postgraduate Medical School a four-page printed form is currently used. The headings on page one, with appropriate space for the information required, are:

Department: ....
Head of Department: Name: ....
Head of Section: ....
Investigator: Name: ....
Brief Title of Project:
Object of Investigation:
Other Departments involved:
Assistants and their responsibilities:

Page two asks several general but pertinent questions, such as:

Patients – How chosen?
Specify how consent is to be obtained?
Controls – How chosen?

Drugs used – Detail of side effects and toxicity:
Has approval of the Committee on the Safety of Medicines been obtained?

Yes/No/N.A.

Radio-active substances:

Has the approval of the Radioisotopes Panel been obtained:

Yes/No/N.A.

RIP number . . . .

Page three requires details of the procedure and here an extra sheet can be added. Page four asks if the investigation has been accepted by the Departmental Research Committee and for any general comments by the investigator, head of section or head of department (It also requires their signatures). The completed form, in triplicate, is inspected by the ethical committee, discussed and, if approved, signed by the Chairman.

One copy is returned to the investigator as a reminder of what he proposed and will be allowed to do; one goes to the head of the department concerned and the other is on file by the administration of the institute where a register of all projects is kept.

There are other constraints. If a new drug is to be used then it must have been approved by the national Committee of Safety of Medicines. For a new procedure or a new drug the Ethical Committee may limit the number of patients to be investigated and ask for a report before allowing continuation. Moreover, no experimental procedure, invasive or non-invasive, should be carried out in "controls" or "subjects" without fully informed and signed consent being obtained. Better still, the subject should be informed in the presence of a witness and the fact recorded in the case records.

Experiments on children present formidable difficulties and some of these have been discussed in an editorial in the Archives of Disease in Childhood (1973, *48*, pp. 751–752). The legal position is far from clear because a parent cannot give consent for a child and the legality of experimentation in this situation has not yet been contested in court.

It is in everyone's interest to review the working of an Ethical Committee from time to time, not just to tighten up its administration but because ethics change and so do public attitudes to research. The function of the committee however remains unchanged, to protect the consumer and contractor, by moral judgements and fairness, to the best of its ability; it must also be fit and able to defend its own decisions. To help individuals to keep up-to-date there is now the "Journal of Medical Ethics" published for The Society for the Study of Medical Ethics, London.

EVIDENCE

Scientific research does not demand reverence, but evidence from its delegates. Indeed, a little irreverence (not irrelevance) is an excellent starting point for any good project. The difficulty then, is collecting evidence and making conclusions.

Before making conclusions it is advisable to examine the data and the logic carefully. What is required is "hard-nosed" evidence, an American expression describing exactly what is wanted. Many people confuse evidence with emotion. The thing to do is to apply the rules of evidence, which differ little from the rules of civil law. A scholarly dissertation on *Evidence in Science* by Kenneth Stone, Emeritus Consultant Physician to the Middlesex Hospital, is well worth reading (1966, Wright, Bristol); it is nicely written and eminently readable.

- All the facts of the investigation should be disclosed. This does not mean that all detail of observations must be included, but it is important to include abnormal results if these are not due to faults of technique.
- The burden of proof and counter-proof lies with the investigator. While there is no such thing as absolute proof in science, the degree of proof needed varies with the type of work done.
- There should be proof beyond reasonable doubt on the balance of probability, which means using statistics intelligently to support the proposed explanation, what is known in Law as "best evidence".
- Circumstancial evidence can be admitted provided that it is supported by documentary evidence from other publications. Circumstancial evidence for an hypothesis may be the best that can be obtained for the moment. It is allowed in law. The big trouble with circumstancial evidence lies in its interpretation; what may be obvious to one is quite obscure to another. It is a two-edged sword which cuts both ways but always tells the truth. Sometimes incorrect interpretations are arrived at simply

because we don't have all the evidence and so efforts must be made to look for supporting facts.

- Hearsay evidence is prohibited. Surmise, intuition and guesswork may be written in the discussion section of a paper for publication, along with personal opinions, but have no place in any conclusions. The testimony of a world authority may have sufficed in the past; today it just will not.
- In interpreting the results which appear to support an hypothesis, arguments for and against should both be adequately presented. Although you may think that the data recorded lead to only one conclusion, another may argue otherwise and set up experiments designed specifically to contradict it. When a judge sums up a case he recapitulates the facts that have been put before the jury and comments on the speeches of counsel. He does not offer new evidence, but he does discuss both arguments, for the prosecution and for the defence. So too with scientific discussion, for when this sifting of fact and fiction is done the evidence becomes clear.

EXHIBITION

An increasingly popular method of display at large medical meetings is the personal project exhibition and the ideal occasion to show your research. Too often this is put together quickly, economically (usually there is no money set aside for such things) and without much imagination. So why bother at all? There are five good reasons why you should.

1. An exhibition affords an unique opportunity to interest the "public" whoever that might be; some for recruitment, others to demonstrate advances in medical research and this is your niche.
2. It provides the chance to illustrate in three dimensions a new slant on everyday things.
3. The artistic emotions may be developed by the beauty in design of the exhibit.
4. It is possible to show working models of how things happen, which can be studied by any observer at leisure.
5. It's a grand opportunity to meet people, talk privately, answer questions and, in general, converse with others who share the same interests. This is no time for wordy descriptions by yourself and not the time for your visitor to have to read a great deal on your exhibit; so keep it short!

SOME POINTS TO NOTE

Most research exhibitions are dull and lifeless. This need not be so, providing the stand is planned well in advance and given considerable thought. It is often worth asking the advice of colleagues by saying, "How would you plan an exhibition to show . . . ?"

- An exhibition must interest the expected visitors. It will probably be in competition with other exhibits for attention, so it must have something special to offer.
- There should be a reasonable budget allowed so that the

exhibit stands a chance of accomplishing what it sets out to do. A clear plan of the lay-out is absolutely vital and should be presented for comment early. In an exhibition, things are displayed and the art is to show them to the best advantage, reflecting the aesthetic appreciation of the person making the arrangement and so stimulating the viewer to action.

● Two dimensional or 3-D effects may be used but elaborate techniques are not required. The exhibit can illustrate a general theme or tell a story in a continuous way (that is, continuity with the minimum of descriptive matter). It is important that the viewer receives a coherent arrangement but it is equally important to try for an original approach.

● Written hand-outs are valuable so that the visitor can, if he wishes, read about the exhibit to refresh his memory later. Clear descriptive writing on one side of a single sheet is suitable.

● Several general features should be decided with the help of someone experienced, either in medical or commercial advertising.

What kind of background? Mirrors will divert light, can show the back of a piece of apparatus and give a sense of depth to the exhibit.

What basic shapes? Angular, straight line, curves?

Texture? Smooth wood, sawn wood, paving stones, plain and patterned wallpaper may all be used with effect. Curtains and other fabrics also help.

What about colour?

Should the basic design be symmetrical or asymmetrical? Will the arrangement be simple or fairly complex?

What kind of lighting? Front, side or back-lighting – each has a different effect on the exhibit. Should lights be coloured? Some exhibits of clear or opaque glass, will transmit light and alter the emphasis of the whole stand. In the same way, a "window" at the back of an exhibit provides depth.

● Try pen and ink sketches of what you want to demonstrate: sometimes large sketches will carry more impact than photographs.

● Additions, such as shells, leaves, branches of trees, flowers, often add "the feminine touch" to round off an otherwise rather mundane exhibit. Pot plants, according to the time of year, give a homely effect.

- Frames and mounts should be considered to give prominence to the particularly important aspects of an exhibit.
- Lettering should be bold, large and at eye-level; it can be flat or raised (using cut-out cork letters) and may be black, white or coloured. Maps and diagrams are valuable presentations.
- Whether the exhibition stand is wall-mounted or free-standing, the use of models (which can be cardboard, paper mobiles or sculpture, of wood or actual apparatus in daily use), photographs and continuous film loops should be considered.

There are new techniques being developed daily in commercial advertising. Some may be adapted to your own purposes, so be on the look out for them.

EXPERIENCE

The Chinese have a proverb which says that "Experience is the comb that Nature provides a man when he is bald". It contains a grain of truth, but no more. True, nothing's cheap these days, least of all experience, but does it count for anything?

Surely, ageing, as with wine improves with time, but the product must be good to start with. The experience of someone who has undertaken a great deal of research in many fields, who has understood what others have done and written (and then did better) is of enormous value for the newcomer to the subject. Bronowski put it quite clearly when he wrote that the real difficulty in research is asking the right question; when you do, the answer may come easily and effortlessly. Experience helps to formulate the right question; it may also ward off asking the wrong ones.

Research is rather like war. Skirmishes are won, not battles; breakthroughs occur, but only on a minor front and certainly not all along the line; no research can eliminate all the variables. In a controlled situation, a lab. experiment for instance, conditions are artificial for there may be 50 factors affecting the status of a patient yet you are merely looking at one isolated instance (The intrinsic elements are numerous and distracting). The more laboratory-like the experiment, the more artificial the situation, the more suspect become the results; the real world is so different. Isolated pieces of information you may have, but not a concerted whole, and this is where an experienced colleague may help enormously. It's what you learn from experience that matters.

EXPERIMENTS

Every surgical operation can be considered an experiment, because the patient is unique and the circumstances different from before. But this is an absurd situation, so what do we mean by an experiment? The dictionary says that it is the "action of trying anything: a test or trial, now archaic" (thank goodness: that lets a surgical operation off the hook) but goes on to state "an operation or action to discover something unknown or to test an hypothesis" (which puts a surgical operation on the hook again). Experimental is defined as "something known from experience", which is not what we have in mind at all. So what do we mean? Trial and error, or real scientific discovery?

An experiment is a contrived situation to allow observations under somewhat artificial conditions. If this is remembered then much valuable information can be obtained (and has been obtained) from the lab. and applied to Man. The general philosophy is perhaps brought out most clearly in Claude Bernard's *An introduction to the study of experimental medicine*, originally published more than a century ago (translation by H. C. Greene, 1957, Dover, New York); a work that should be read and reread by everyone.

Three aspects have to be considered:

- Experimental design: several good books on that subject including those of D. R. Cox (1958, *Planning of experiments*. Wiley, New York) and R. A. Fisher (*The design of experiments* – 6th edition, 1951 – Oliver and Boyd, Edinburgh).
- Experimental method, largely how the work will be done (see METHODS and METHODOLOGY).
- Experimental data, how this will be handled. Many people believe data to be the end result of an experiment. This is not so. Data are of three kinds, raw data, informative data and interpreted data (the only kind worth reporting).

FALLACIES OF LOGIC IN MAKING CONCLUSIONS

A fallacy is any sort of mistake in reasoning or inference which causes an argument to go wrong. For instance, three Irishmen went to three pubs. At the first they drank whisky and water, at the second brandy and water, at the third vodka and water. Next morning they came to the conclusion that the cause of their hang-over was from drinking water! Moreover, a modern hospital measuring the output of urine would have supported this hypothesis. Errors can be explained as part of human fallibility: there is nothing mysterious about them in a scientific investigation. They are numerous. To give examples from published papers would almost certainly be libellous; hence they are only listed here.

- Ambiguity, that is, using terms with more than one meaning, or when each word is correct but the sense of the whole sentence is not clear.
- Contextual fallacies, by taking words out of context.
- Fallacies of "significance" and incorrect emphasis of words in a sentence ("*All* were cured except three patients with. . . .": *all* is not true).
- Words directed against an author to show that what he said cannot be true. Read the Lancet of 100 years ago when people were more outspoken, but even now the same thing occurs, although often wrapped up in woolly words.
- Arguing from authority. A statement cannot be proved true or false just because someone else says so. The pragmatist concludes that an hypothesis is true if it works in practice; this is not so, although on the whole science accepts things that work and rejects those that do not (the same in technology).
- Arguments appealing to sentiment: no statement can be shown to be true or false just because that is the way people feel about it. Hence controlled clinical trials – and there are too few of them – in assessing any new therapy.

- Ignorance. A Statement cannot be accepted as true or false just because there is no evidence to disprove it. Positive evidence is always required.
- Begging the question, where a conclusion is one of the premises. This occurs commonly where one of the premises could not be known to be true unless the conclusion assumed it to be true.
- The fallacy of composition: what is assumed to be true of a part must be true of the whole.
- The fallacy of division: what is true of a whole must be true of all its parts.
- The irrelevant conclusion. The investigator starts out to prove one thing, but then proves something else.
- The non-sequitur (or it does not necessarily follow that. . . .) One of the commonest errors in medicine.
- Statistical fallacies, that is, by misusing statistics. It is important to know what methods were applied and how variables were controlled. Absolute falsification in making conclusions rarely occurs. Since nearly all observations depend on preconceptions, why should conclusions be exempt?

FILING & FORMS

Filing is a personal exercise, carried out more efficiently by your secretary (if you have one) because she has been trained in the art. There are usually several good books on the subject of office organisation in most public libraries and chapter 20 of Morton, 1974, *Use of Medical Literature* (Butterworths, London) by D. C. Roberts outlines filing for medical people. A paper to read is that by McClelland (1970. Brit. J. Anaesth. 42. 744) on information retrieval. Essentially, there are four methods for the researcher:

- to use a box file and put everything in the order in which references, publications and notes are received. When the box is full, buy another. This method is frowned upon by those with tidy minds, but it does have a certain appeal, especially for the man writing a thesis, because of its lack of orderliness. Sorting through what has been acquired, a useful exercise in itself (at Christmas perhaps, to escape those dreary children's parties), can be a revelation and a stimulus to new thought.
- to use a notebook for essential references, but again quick retrieval of information is difficult and often impossible when you have several notebooks. This method should not be confused with the "little black book" carried constantly to jot down ideas, where part of the discipline is to look through the entire contents at frequent intervals.
- the card index file has been discussed under "Reading" and there is little more to add. As cards accumulate some form of classification, with coloured separating cards for subject headings, becomes imperative. How cards are classified matters little but it is wise to plan a system which will still be convenient when the collection has increased ten-fold. Cross reference cards will also be required, a time-consuming and rather boring chore.
- one can use cards with holes punched all around the edges, a notch being cut opposite the subject noted on the card. These

were popular at one time and certainly are fun to play with; a knitting needle is threaded through all cards at the appropriate hole and the lot lifted out; those remaining (having been notched) are the references you desire. Cards do not have to be replaced in correct order in the container after use which saves a little time, but the possible variety of subjects is strictly limited by the number of holes available (about 60).

Whatever method is used, be sure to lock up the data; there is no point in giving the snooper a chance to be one ahead.

Much research involves recording information in a routine manner and for this a special form may have to be devised, for speed and to insure that important items are not omitted. With a standard form, analysing the data may be greatly simplified. Some thought has to be given to the design of the forms (and consult the Civil Service publication "The Design of Forms in Government Departments", 1971). Data sheets and forms should:

● be attractive in appearance, reflecting a good design, and be functionally effective (that is well thought out).
● use diagrams where these are self-explanatory (for instance if blood pressure is being recorded, mannikin figures – lying, standing, walking and running – will indicate more easily than words what is wanted.).
● have minimal text.
● explain what is required at the *top* of the form, before allowing space for recorded data.
● ask unequivocal questions so that the answer is yes or no.

FIRST PUBLICATION

For the young researcher preparing his first paper for publication, a lot may seem to be at stake. The subject may concern the results of careful investigations made over several months, or even years, the impact of which can be ruined by bad presentation. His chances of promotion, if not actually dependant on his paper may at least seem so to him. His future and reputation as a serious investigator may be thought to depend on what is written how it is set out and the way he develops a sound argument in discussion. It often does!

The late Cyril Connolly, author and journalist, once said, "The only point of writing is to write a masterpiece". This is an ideal higher than attainable for most of us, yet it is the point to aim at. The standards set at the start can usually be maintained later; hence, the importance of style and exact writing for that first paper. Improvement in writing comes with practice.

FREEDOM

In a world where freedom of the individual to go his own way is being eroded by the State, it is refreshing to realise that, in most countries at least, the researcher still enjoys four freedoms:

- To write and say what he likes – but supporting evidence is demanded. Even if his interpretation of the evidence is incorrect, there is no rancour.
- To choose what subject to investigate – but the reasons for such are expected.
- To quote published papers without redress – but the actual words should be quoted in full, even when a different interpretation is to be put on them.
- To apply for financial support for the project – but you will be expected to account for your actions and show a return on the outlay; value for money is simple justice. If you should think that a grant-giving body owes you a living, the sooner you find a job elsewhere the better.

The price of freedom is eternal vigilance, as every scientist knows. Many have suffered for just that and the newcomer should be made aware of exactly what is at stake so that these freedoms are never abused. I would also add another to this short list, the freedom to make mistakes and to learn from them; abuse that freedom at your peril.

GAMESMANSHIP

Gamesmanship, a term coined by Stephen Potter in 1935, means putting "one over" the other fellow (a "ploy"). It is a great spur to research, for it adds a special flavour and a sense of competition to the work. Everyone has his own collection for writing and lecturing, but here are some to start with:

- Refer to unpublished work; it cannot easily be contradicted and indicates that you have special inside knowledge.
- Refer to "personal communications" from others, even if they are only chance remarks.
- Quote the Japanese, Russian and Chinese literature especially if there is no summary in English.
- Name dropping, especially if a Peer of the Realm, is always useful.
- Refer to rare journals, particularly those that are now out of print.
- Refer to a long forgotten lecture by a notable person now dead.
- Quotations from the works of Plato, Sophocles and Virgil are always difficult to check. Avoid, however, quoting in the original language because it may be misprinted and a classical scholar will subsequently point this out so that the ploy is lost.
- Quoting Government Acts is quite useful because amendments made later may have only changed a word here and there. Moreover, it is hard work going back to the original Act of 1896 (which may have dealt with the relief of Mafeking anyway). In a list of references, a series of Arabic and Roman numerals, brackets, capital and lower case letters, always look good to demonstrate your own erudition.
- Snide remarks about the position of a previous author are still allowable but have to be well wrapped up to avoid a charge of libel:

"Smith, a general physician, found . . . , but a modern vascular surgeon would dispute his observations."

- On a personal note if your name is Herbert Smith, change it to Hensley, Harvey or even Hackforth Smith. Soon people will begin to think your name is hyphenated. There's a certain amount of snobbery in all of us and, for Britain, a hypenated name carries a little of nobility about it. In the U.S.A. I suppose "Lord" and "Earl" help equally.
- Another is to make up new words for quite simple things, but to a large extent the psychiatrists and psychologists have beaten us to it. Even so, the word "cybernometer" should confuse most people (for phrases see JARGON).
- Give references, almost exclusively in a publication, to papers of your own, friends and important people (the art of arse-scraping).
- Using histograms with heavy shading in favour of the author's treatment when a simple table of results would show nothing significant.
- Giving placebos and not the best available drug currently available, to the control group in a clinical trial, so that the new drug looks better on performance.

For a smooth effect, new ploys must be rehearsed well beforehand. The natural "con man" will develop a large repertoire with little effort, but, for most of us, hard practice really is necessary. Richard Asher's book has a chapter on "Medical Salesmanship" for further ideas.

GRANTS AND GRANTSMANSHIP

Research costs money. So where does the money come from and how do you obtain your share? There are 5 sources:

- Your own pocket, the quickest way to bancruptcy.
- From local funds of the institution, which may suffice to get the work started but rarely allow you to finish in style.
- Central government funds, such as the Medical Research Council and the Department of Health and Social Security in Britain.
- Grant-giving bodies, such as the Wellcome Trust, the Nuffield Foundation, the British Heart Association and a host of others to be found in a little booklet issued by the British Medical Association (*Research Funds Guide*, 1968).
- From Pharmaceutical companies who may wish to obtain information about one of their new products. Often such financial grants (for the work involved and never as a personal retainer) are for quite short periods, perhaps 6–12 months. For 'on-going' research such help is valuable. The big risk is that the researcher will try to please the company in order to justify the expenditure. Hence, the results may be interpreted in a biased manner which shows in the 'discussion' section of the publication. It is a real risk and should be discussed with the representative of the pharmaceutical company quite openly before any grant is accepted. The freedom to publish, whether the result of investigation is good or bad, must be a condition of acceptance. Most companies willingly agree, but from experience disasters can happen. In the end, of course, inaccurate information is to no one's credit.

The office of Health Economies information sheet no. 25, "Expenditure on Medical Research" (in Britain) 1974, noted that expenditure on research from all sources rose from £26 millions in 1962 to £108 millions in 1972, a four-fold increase. Government expenditure, however, has begun to change in character since the Rothchild report introduced the "customer-contractor' principle in 1971. (The Lord Privy Seal. *A framework for Government Research and Development*. Cmnd. 4814. London, H.M.S.O.) As an editorial in the *British Medical Journal* (29th June, 1974) noted, it is debateable whether one can really draw up a list of priorities in research. "Should they be determined by the size of the groups of sufferers? Or by the calculations about the likelihood of making a research breakthrough?" Clearly there are no self-evident answers to such questions, but even so, a researcher has to apply for funds and do it properly; he cannot afford half-measures.

APPLYING FOR RESEARCH GRANTS

Rightly or wrongly, a great many junior doctors believe that they have to do research in order to obtain an interesting senior appointment. There will always be some expert around to advise you otherwise, that your job is to treat patients, not to live in a lab. It is not intended to argue either way, but I would point out that research has an educational function, for education is the accumulation of experience, not facts (Welbourn, 1966). Sooner or later, it will be necessary to apply for finances to support the research project. It is never too soon to apply, for it may take 3–6 months to succeed. How is one to go about it?

Doris Merritt (1963) has described the application for funds as "grantmanship: an exercise in lucid presentation" and Allen (1960) has recorded some of the reasons why grants are not approved, and these are usually self-inflicted wounds. An application for funds, like an application for anything else, should have a certain structure and provide information clearly and concisely. Writing the application will require as much thought, and nearly as much time, as a paper for publication. The following format is proposed with each heading starting on a new page unless the donors, such as the Medical Research Council, have their own form to be completed.

- *A TITLE PAGE* which should carry a short descriptive title, the applicant's name and address, and the date of submission.
- *THE INTRODUCTION* should provide a very brief summary of the problem it is proposed to tackle, and only a few relevant references to previous work need be given. A good method

is to pose questions which are of importance and which you seek to answer. These can, and perhaps should, be provocative. They are likely to interest the reader and persuade him that the application is important.

- *METHODS AND MATERIALS (or PLAN OF THE INVESTIGATION)* The description of both should be explicit. It requires considerable judgement to know how much to include, but the application should contain sufficient detail to allow another to carry out the work as planned. If analytical methods are well-known then a single reference to a published paper is all that is required. If, however, new methods are introduced they should be defined in detail; if they are long and complex it may be more convenient to place them at the end, as an appendix. One has to explain complexities in terms which will be recognised by the reader and in words which permit of only one interpretation. You have to demonstrate by irrefutable logic and imaginative insight that there is a gap in knowledge to be filled.

If it is likely that one method may succeed and another fail, then alternative methods should be included. The successful applications are those which contain a well-defined problem and a well-defined approach. In applying for a research grant the only saleable commodity is a good plan to acquire knowledge which can be used by others; better still, support that exciting wild idea by some experimental evidence, a pilot trial, or a very convincing argument. If pilot studies have been carried out, showing that the work is feasible, they should be recorded with all the preliminary data.

- *FACILITIES AVAILABLE* These should be described accurately. If certain equipment is required it should be intimated and its cost included in the budget. It is also advisable to mention the names of experts who are available for consultation or technical help.
- *WHERE THE WORK WILL BE DONE* should always be specified. If certain measurements are to be made at another institute this must be stated and the consent of any collaborator mentioned.
- *THE CURRICULUM VITAE* of the chief investigator can be included here, or as an appendix, and will also contain pertinent previous publications. This section is important for by it the assessor may be able to judge the background of the applicant

and his potential ability in research.

● *FINANCES REQUIRED* These should be set out as:

A Capital costs

B Recurrent expenses

It is unrealistic not to ask for what is required to complete the job. One should ask for finances to cover the time that will be devoted to the research. In the U.K. a three-year limit is commonly enforced, so if the research is likely to take 5 years it may be worth dividing it into two projects, the second to be supported when the first is finished. The cost of personnel, animals and their husbandry, materials and other disposable items, should be accurate.

● *REFERENCES* must be relevant and recent. You should have read them and judged their importance and applicability. If they do not support your proposed work, do not include them.

The successful suppliant will have set out clearly, not perhaps in so many words but crystal clear nonetheless, the answer to four questions:

Why do I want to do this?	– the need
What do I want to do?	– the plan
How am I going to do it?	– the methods
What do I need?	– the resources.

It's worth re-reading the application with these four in mind as a final assessment before sending it off. If the answers are not clear to you, then how can you expect them to be clear to others? Ambiguity is the great stumbling block in any research petition, so avoid it; it cannot be stressed too often and too forcibly.

All applications to grant-giving bodies are referred to a consultant, or other expert in the field for his advice. He will examine your application and ask 5 questions, the answers to which should be obvious from your writing.

1. Is it important?
2. Can it be done?
3. Is the investigator competent to carry out the work?
4. Can it be done within the specified time?
5. Are the costs realistic?

Applications are turned down commonly because:

● They have little scientific merit.

- The work is based on an unsound hypothesis; is more complex than the investigator realised; is premature or the description is too nebulous, diffuse and obscure. The error here is in not taking sufficient trouble and time to think about the project, or just plain ignorance.
- Excessive reliance on others to carry out the major part of the work and general lack of common sense on the part of the investigator.

There is another side to the coin, the lack of ready cash for the committee to distribute. So look at the grant-giving body's assets – shares, public donations, profits – and plan the demand accordingly. Every time there is an industrial strike or a major disaster, someone loses, someone gains; perhaps monies have already been allocated for a building project which fails yet have to be spent within the time allocated, but cannot, so put out your hand.

HIDEAWAY

A notice on the door "PLEASE DO NOT DISTURB" merely tells others that you are in your office and therefore available: it does not guarantee peace, often only a queue of people waiting to talk to you.

A desk is needed to write on, not to hide behind, so what is meant by Hideaway?

If you want to think, hide. No one can work out the details of a research project with telephones jingling, interruptions from passers-by, unwelcome visitors. Where you hide matters little, the important thing is that you cannot be found. Hence, the public library and even the toilet may provide a refuge.

At the Royal Postgraduate Medical School there are carousels beneath the library, accessible only by a winding staircase. Peace and quiet are assured for it is possible to dive below to spend a couple of hours writing or doodling without anyone knowing. It gets a bit lonely at times, but then you can come up for a break when some work has been done.

Everyone in research needs a refuge; choose it with care and don't tell anyone, not even your secretary.

HISTORY

Henry Ford said "History is bunk" but he didn't mean it. History looks back and only occasionally allows one to predict the future, with a high probability of error. Even so, history is useful in research in two ways:

1. to see how the development of your own subject has progressed. Unfortunately history, because historians are unable to write in perspective, will not tell you what has happened in the last 20 years, a period when you may already know that the subject has expanded very rapidly.
2. to realise that your new invention was antedated by a century or more, even though the methods may have been crude and inadequate, yet the conclusions were correct. MacFarlane (1974. *Thrombos. Diathes. haemorrh* (Stuttg.) *32*, 1–8) has questioned the connotation "conclusions" and suggested that often these represent an hypothesis, no more; if so, there is encouragement to repeat the work and progress a stage further. Reading history cuts us all down to size and makes us realise just how much we owe to former generations (some investigators lost their positions and even a life or two in the pursuit of truth, a thing unlikely today). It provides an essential perspective view.

Try to read such books as:

- *Medical Bibliography* (1970) by F. H. Garrison and L. T. Morton. Third Edition (Deutsch, London, 872 pages)
- *Modern methods in the history of medicine* (1971) by E. Clarke (Athlone Press, London, 389 pages)
- *A history of medicine* (1947) by A. Castiglioni. 2nd Edition of 1146 pages. (Knopf, New York) – a scholarly work.

All these make profitable reading while sunbathing on a beach and will also provide ideas for a stimulating introduction to a lecture or publication on the subject of choice; at least you might be recognised as a well-read person. Then delve into the medical histories of other countries. A good list is provided by E. Gaskell (Chapter 19 in L. T. Morton's (1974) *Use of Medical Literature* Butterworths, London) which also includes histories of specialist subjects.

● Henry Sigerists' *The great doctors* (Dover Publications, New York, 1971) in 400 pages and L. S. Kings *A history of medicine* in 302 pages (Penguin, Harmondsworth) are easily portable.

HUMOUR

Humour is an essential ingredient for successful team work. Not the bawdy kind but the real humour, the ability to laugh at yourself and laugh with others. It's a subtle form of human relationship which alone will prevent internecine disputes and petty jealousies.

Research is an emotional subject. We rejoice when things go well, are depressed when they go badly. So there's plenty of scope for humour. Indeed, humour is necessary for sanity, success and comfort.

HYPOTHESIS

The Shorter Oxford English Dictionary defines an "hypothesis" as a proposition or principle stated merely as a basis for reasoning or argument, or as a premiss from which to draw a conclusion; and, a supposition or conjecture put forth to account for known facts – the generally accepted definition in science. An hypothesis is a mere assumption or guess, no more and no less and frequently an idea snatched out of thin air: it is imagination from within (and see SCIENTIFIC METHOD).

An hypothesis requires support from research and so must be stated formally and specifically. For this reason alone, much research in psychiatry is unsatisfactory: the hypothesis is poorly defined in psycho-analytical language, often in bad grammar and in peculiar English too.

There are five requisites for a useful (as opposed to a useless) hypothesis, for it should:

- explain the known facts,
- be consistent with all the known facts,
- be no more complex than necessary,
- aid prediction of new facts and relationships, that is, to explain more than it was designed to do (its extra-mural implications),
- be susceptible of verification or refutation: its testability. Karl Popper believed that the function of an hypothesis was to provide clear statements that were susceptible of disproof (by observation, experiment or both).

The supporting evidence may be presented as graphs, tables and simple description. In science the object is precise measurement. Galileo, whose hypothesis was that the earth moved round the sun and not vice versa, said, "measure what is measurable and make measurable what is not". In the biological sciences we know that all measurements are inexact and the only way to describe this variability is to use statistics intelligently.

In medicine there are two concepts of diagnosis. The inductive simply observes and collects information, on the whole degrading to the clinician who becomes no more than a computer, as Medawar once pointed out. The other, the hypothetico-deductive, allows an hypothesis based on observations: the provisional diagnosis can be rejected or accepted, by further observations, usually specific tests. Cure of the disease validates the hypothesis.

An hypothesis, then, is based largely on imagination and so liable to error simply because of human fallibility; we all make mistakes, the commonest in believing that our observations are unbiased. The idea of unprejudiced observation is a philosophical myth, as much as the belief held in the last century that all truth lies in nature and only requires observation to be discovered. All observations are based on preconceptions.

Scientific procedures and technology have their part to play in research: on the whole they are boring compared with the intellectual stimulus in formulating an hypothesis. Speculation is the breath of life in an hypothesis, for it comes straight out of the head of the investigator.

IDIOSYNCRACIES

As a research man or woman you will be looked upon as a strange person by many. So, give them good value by developing some strange habits: they are useful especially when abroad. Denis Melrose, when he went to Russia in 1966 with his new heart-lung machine, always had a fresh English rose in the operating theatre. In those early days of complicated heart surgery perhaps it was needed for morale and the patient's welfare. It became a legend and was copied by many others almost as an act of faith. The early machinery is forgotten but the rose is remembered long after that historic visit.

INCENTIVES

The old adage that "you can take a horse to the water but cannot make him drink" applies as much to research as it does to industry. The solution, surely, is to devise incentives. Here are four:

- The carrot, which works for the ass (or so we are told). This is really expectation, of one's name on a published paper, a successful thesis for a higher degree, of future prospects either by promotion within the research team or the eventuality of setting up the candidates own laboratory.
- The fact that an individual may be working in a congenial environment with others, perhaps for the first time, may suffice. Regretably, this does not last long and we have to move on to the third incentive.
- Job satisfaction, a modern term for a simple request that everyone should enjoy what he does, which demands that the job be varied and interesting.
- Money: McCormick's Law states that, whatever the salary, take-home pay covers only the bare necessities of the individual's style of living. Hence, raising a person's salary only increases the proportion gathered by the tax-man and does not make things easier. One way is to pay expenses to conferences (preferably in an equitable climate), the other is to provide "perks" but these are more difficult to devise.

INTEGRITY

However debased his fellow citizens may become, the researcher cannot afford to "cook the books". If he does he will be found out and exposed. This is the one fundamental demand in research, to be honest at all costs; there is no cloak for the wicked. Hence politics and personalities cannot safely be served except at the risk of your own reputation. Which would you prefer to lose? Patronage or reputation? The choice is always loaded for the man who sincerely believes in what he is doing.

INVIOLABLE RULES IN SPEAKING ABOUT YOUR RESEARCH

When you are asked to talk about your research there are 3 inviolable rules which should never, never, be broken:

- Keep within the allotted time, which requires practice beforehand and a great deal of self-discipline. If the audience is unsatiated by your remarks they can always ask for more, but they rarely do.
- Make it interesting to others who may not really understand what you are doing. This demands thought, careful planning and much personal effort.
- Give the meat only and leave out all the trimmings, which needs good judgement.

A talk bears no relationship to the paper you will publish on the subject. Indeed, if some of the audience come to realise later that these were two different offerings, so much the better.

JARGON

What is jargon? The *Concise Oxford Dictionary* defines jargon as "Unintelligible words, barbarous or debased language". When one is told to avoid jargon in writing (and why not in speech too?) commonly the advice is to avoid a mode of communication full of unfamiliar terms. This is the nub. To a doctor, a technical report in engineering appears to be full of jargon; the reverse is also true. We have therefore to distinguish between "gibberish" and a "technical vocabulary". Elder (1954) argued that if the technical vocabulary of a science is jargon then all such authors must be condemned, for in no other way can they achieve clarity and conciseness of expression in writing. But what is the use of fine research if the results cannot be communicated to all who might benefit? Publication is the end product of research, for without it research becomes sterile.

In medicine, technical terms have been defined and are recorded in special dictionaries. These words have been constructed and designed to serve the writer's needs with precision and economy. Some words, of course, have different shades of meaning and, when used in statements, their precise meaning should be explained. This implies that simple words should be substituted for the technical term where possible; they should, as a courtesy to the reader, for the object is not to produce an orgy of difficult words for economy, but to write clearly. Research has now become involved with "management", a specialty where confused jargon is so common that one wonders how managers talk to each other.

Jargon of quite a new and frightening kind has entered medicine and allied subjects. Most of the new words are so joined together that the sense is always difficult to discover and often quite meaningless. This is a great pity because administration needs a simple language, not a foreign tongue. If you really wish to confuse, then think of a 3 figure number and string together the words corresponding to each digit in each of the three columns of the "Jargon Chart" to produce the required phrase. For instance, 555, (synchronised incremental integration – or "togetherness" in normal language) and 008 (overall management policy) are useful at conferences.

The Jargon Chart

0	overall	0	management	0	aversion
1	anticipated	1	logistical	1	expectation
2	systematized	2	reciprocal	2	capability
3	functional	3	transitional	3	flexibility
4	responsible	4	organisational	4	concept
5	synchronised	5	incremental	5	integration
6	projected	6	limited	6	implementation
7	balanced	7	marginal	7	ideology
8	compatible	8	material	8	policy
9	optimal	9	oriented	9	resources

JOB DESCRIPTIONS

Beloved by the administration of Industry and State, job descriptions are worse than useless in research; they are restrictive. The nine-to-five mentality has never produced anything worthwhile, except a good attendance record and that is quite definitely not research. This is not to say that no discipline, no direction, should be countenanced.

One of the best "job descriptions" is contained in the University of London contract for a personal chair in Surgery. "The professor shall teach and not indulge in private practice" – that's all there is to it: nothing about hours, number of patients or publications, and as a result most incumbents work hard for long hours enjoying the freedom from restrictions imposed by a benevolent university. How wise!

Researchers should not, however, be unaware of what is required. Quite the contrary! If the post has to be surrounded by rules and restrictions, then it probably has little to do with research: you want a lap-dog and have to pay the price, but do not expect new ideas and certainly no spark of inventiveness.

JOURNAL

Which journal shall have the privilege of publishing your gem? The matter requires considerable thought. For one thing, for whom are you writing? To whom do you want your message to go? Do you want to receive comments on the work? – if so, the paper must be acceptable to a journal with a correspondence column. Of course, as a last resort it may end up in the British Journal of Pornography, widely read, but not by those people you wish to influence: you may build a reputation in your subject, but in the wrong quarter. The choice of Journal depends on:

- The general quality of your paper and the prestige of the journal in which you wish to publish it.
- The size and circulation of the journal. What public do you desire to reach? Will the paper be of interest generally or is it intended for the specialist? Does the journal have editorial and correspondence sections? Will these be important to you?
- Speed of publication: is this important? How long is the waiting list for the journal in which you would like to place it? Some journals print the date of submission as a footnote to each paper and by looking at several it is easy to calculate the average waiting time; with weekly journals it is usually short, but in specialist journals it may be over one year.
- Your illustrations may require good quality paper for accurate reproduction which will necessarily limit the choice.
- Examine the journal for the type of material selected. Fashions change and some journals are pleased to expand their range of subjects to attract a larger reading public.
- Read the editorial notices, note the editorial committee and the directions about submission. Familiarity with the lay-out of papers, how the references are set out and how long the summary should be, are clearly important. Is the summary at the beginning or at the end of papers?

You will have to ask some fairly searching questions, such as whether the work is original in your own country and whether it is sufficiently important for the journal of choice. When it has been written, it is advisable to consider the choice of journal again, as well as the soundness of the statistics, the logic and the quality of writing.

Most journals devote a page or two for guidance to intending authors. These should always be consulted before the final typing of every paper, vague though the notes of guidance may be. *Clinical Science and Molecular Medicine*, 1974 (volume 47; pp 1–XV11 "Guidance for authors") sets out its policy in detail and includes notes on S.I. units, chemical terms and accepted abbreviations. *The Journal of Physiology*, London (1966, "Suggestions to authors" volume 182, pp. 1–33), *The Royal Society* ("Instructions to authors") and a more detailed booklet (*The manuscript* 1969, 6th Edition, by W. Jaggi and issued by Karger, Basel) are all worth consulting.

LEADERSHIP

The role of the researcher has increased and continues to increase in complexity and importance, with a rapidity and to an extent which would have been incomprehensible to Claude Bernard who laid the foundations in the last century.

The person who realises this and determines to do well while keeping ahead of the times is likely to be a natural leader. In research, as in doctoring, leadership is something special, something valuable, which contributes greatly to the output.

Leadership has nothing to do with status. Most leaders are unrecognised until something goes wrong and brings out these special qualities – knowledge, courage, decisiveness, dependability, initiative, tact, justice, enthusiasm and humour, judgement and common sense, tenacity and self-control. A tall order if you intend to be the boss and the leader of your research team (they are not identical). Perhaps the supreme quality in a leader is confidence.

Leadership is a lost art. When a good leader is present, nobody notices; when absent, everyone knows. Besides, his people consistently turn out better research and obtain better jobs. Leadership is essential to an active research team – to direct its energies effectively in the right direction. We now know that the art of leadership can be learned, it is not inborn as a natural gift. We also know that there are four distinct elements involved:

- the characteristics of the leader; these are recorded above,
- the attitudes, the needs and the other personal qualities of the followers,
- the type of research organisation, its purpose, structure and the nature of the tasks it sets out to perform,
- the social, economic and political environment.

All four are important because it means that leadership is not the property of one individual, but a complex relationship among all these variables. Thus, in a successful research team the leadership will change as the projects change – and so will the authorship of proposed publications – for these influences act in subtle and unnoticed ways.

Mutual confidence between the leader and the led is essential: there are many ways of developing and maintaining this confidence – equality of status, ability and energy are obviously important. Hence the profound importance of the construction and constitution of the research team, the great influence of information upwards and the great failure of authority wielded downwards – all so evident in dissident industry today.

You cannot begin to understand people until you know their interests. Most people like to talk about their hobbies – painting, poetry, photography, singing; they are amateurs but experts, and this is seen so clearly in a hospital ward at Christmas time when hidden talents are displayed to the amazement of all.

Not every individual will become a leader. Some are incapable, some have no wish, but all should be given the opportunity to develop their own resources (often with unpredictable results). Leadership at any level is a precious thing; it should be recognised and rewarded.

Leadership ensures good morale. Poor morale is often due to not caring for others, caring too much, and insincerity. To be successful and sustain good morale, the leader must:

- Be affectionate to every member of the team. He must like them all, or get rid of those he doesn't. There are no half-measures.
- Inject respect into every individual, for himself and for the leader – for anyone who wants that exhalted position has to be admired, right or wrong.
- Be ambitious by setting his sights high. Hence, he will not only have to make the work interesting but set a good example too.
- Be active, and so emanate energy in time keeping, in planning, in directing, in doing.
- Be accessible by letting people see you – so stroll around, know peoples' names and what they are doing, appear unexpectedly, talk to others, show interest and just be human. On

my own office door I have a "Do not disturb" notice, which when hung is interpreted as "Now is a good time to have a chat without being disturbed".

- Be articulate by letting everyone know what is going on.
- Be appreciative, be grateful and show it; you will then have more to be grateful for. Spontaneous praise, like humour, has a freshness all of its own, especially when unexpected, which is recognised as solid gold, the real thing.

You may consider this a tall order for a leader: to have to do your homework (instead of watching T.V.), to know everything about your subject. Yes, indeed, but without that you can have no confidence, no conviction, no certainty. Let's consider the alternatives, for you will fail in your job by:

1. Giving no leadership: lost sheep,
2. Employing the wrong people,
3. Rewarding the ineffective and ineffectual for the wrong kind of service,
4. Communicating inadequately,
5. Being inconsistent, punishing excessively, encouraging informers,
6. Leaving jobs and responsibilities undefined and unobserved.

In short, chaos and non-productive research. The choice is yours.

All members of the team must have a common objective, usually to conclude the project. All must consider it worthwhile; all must believe it can be attained. Therefore, there must be a logical chain of command so that each knows who is responsible for what. In this, communication is the keystone, for all must know what is going on; leadership requires well thought out ideas, clearly expressed in language all will understand.

If the key to good management is delegation, and I believe this to be as true in research as in any other occupation, then there should be:

- no delegation without responsibility
- no responsibility without representation.

Both are vital to the making of an efficient team. Participation is the strongest incentive to work well, but it is important to make sure that people know their job well and what they have to do.

Poor leadership is shown by low morale of the team – resignations, accidents, sickness, grievances and complaints, as well as slap-dash work. Morale is related to job satisfaction. Yet the effectiveness of a supervisor in planning, organising, staffing, directing, and controlling has an important consequence on the behaviour of the research team. Organising involves two aspects: the assembly of resources, and the delegation of authority and responsibility. The wise leader will never try to perform the function of subordinates, but a working supervisor will always have more power over them than a non-working one. Success is shown by high morale.

Anyone with integrity who has social and moral values and who recognises the needs of others is likely to be an effective leader. There are many other qualities of leadership – knowledge, courage, decisiveness, dependability, initiative, tact, justice, enthusiasm and humour, judgement and common sense, tenacity and self-control – but even so, leadership without the ability to communicate clearly has little substance and little value. To be able to do something as well as others is not enough: it does not give the leadership position without which a research team is doomed. Only excellence earns impressive results. The innovator, the leader, has special responsibilities to others.

LECTURES AND LECTURING

All too soon you will be asked to lecture about the results of your research. For convenience the word "lecture" is preferable to "communication" and certainly better than "reading a paper": never read, only speak. At first you may be asked to say a few words at a departmental meeting, then 10 minutes in a full programme for professional researchers, eventually a one hour lecture as a guest speaker on your subject. Unfortunately, requests exceed the rate of discovery and somehow you will have to compensate for the disparity; no matter, we all have to find a way, but try never to repeat exactly the same lecture.

Lecturing is more formal than simple conversation but the same advice applies – be natural, be yourself and not some strange animal who suddenly acquires a new language (foreign to your listeners), a new affectation (irritating to most) and a new poise (of unacceptable authority). It is strange how some people change like a cameleon when they have to explain their daily lives – and can only do so in the most complicated terms.

When I went to school there was a class called "Rhetoric". It was an old-fashioned school and the class was the equivalent of today's "top form". Rhetoric means "speaking" although the Shorter Oxford Dictionary includes: "the art of using language so as to persuade or influence others; the body of rules to be observed by a speaker or writer in order that he may express himself with eloquence." Indeed, this says it all – elegance in speech.

The advice given on "Communication" applies and so is briefly repeated here:

- English is two languages (spoken and written) but here verbal expressions only are our concern.
- The paper for publication may be composed piece-meal but the lecture has to be produced as a unified whole. The structure for a lecture is quite simple – begin, tell them your message and

finish. More specifically:

a. Introduction, of yourself and your subject
b. Main message, what you did and what you found
c. Conclusions from your own work.

● The amount of preparation required varies with the subject and the occasion. A good working rule is to devote one hour for every one minute of speaking time to rehearsal and final polishing. Hence, a 10 minute lecture will require 10 hours of vocal preparation plus the time required to compose the script and illustrations. The art of speaking well can be learned (see Calnan and Barabas, 1972) but it is essential to have something important to say and to say it clearly and cleanly. No book provides more than a guide, for success comes only with practice and a lot of hard work.

All too often, much time is spent on preparation of the main message; this is foolish. The very reason you have been invited to talk is because you are an expert in that particular subject; the decision you have to make is what to put in and what to leave out. Composing the main theme involves clarifying, emphasising, condensing, isolating, adding, subtracting, rearranging, improving, selecting or rejecting ideas so that the components of the subject are made on the basis of choice. The essence of composition is organisation, which implies order, and order demands a strictly logical arrangement. Indeed, we can write a simple time sheet:

	Time as a percentage of that required for	
	Delivery	Preparation
a. Introduction	20%	80%
b. Main message	75%	10%
c. Conclusions	5%	10%

Planning and preparing the introduction is by far the most tedious chore because you will be introducing yourself and your subject to the audience while capturing their interest. The conclusions should be self-evident and so require little time to spell out.

● Interest as many of the audience as possible because research is often dull stuff to talk about even though exciting to do; your subject may be rather boring unless you can bring some new quality to bear – so be an entertainer, not just an educator. A lecture and a published paper differ markedly because the lecture is made on demand to a known audience; there may be little time to prepare and even less to rehearse, but any omissions or misunderstandings can be tidied up at question time. The published paper is produced at leisure for a largely unknown audience of readers who can go back over difficult passages in their own time. Enthusiasm in a lecture usually pays off: in a paper, rarely so.

● A lecture should never contain all that you know on the subject. Moreover, leaving something out allows others to ask questions without having to think too hard. The man who puts everything into his lecture is repaid in his own coin, by silence when question time comes around. The fear that others will not think you knowledgeable if you omit something is naive; more likely you will be considered stodgy, unimaginative, over-anxious and dull (which you almost certainly are!) Always allow time for questions afterwards because the cut and thrust of debate may be remembered when your message has long been forgotten.

● There are three things to aim for in a lecture:

a. to reach the minds of your listeners – the intellectual content

b. to make an impression on their sense of morality – what is right and good, or wrong and bad

c. to appeal to the emotions (not the same as an emotional appeal, for remember that we all are emotional, some more than others)

You must give the audience a little of yourself as well as your subject – your own zeal, conviction and energy, even though you are not a professional speaker. Simple recitation of the facts and their interpretation will not be remembered – they are better read in a journal. No-one is interested in how you got there, only in what you found: so tell them what they want to hear.

The important thing in a lecture is to project yourself. The Welsh lilt, the Scots' crispness, the Irish brogue come through in speech but are completely lost in the written paper. What a pity that the personality of the author cannot be proclaimed!

LETTER TO THE EDITOR

If your research has revealed some devastating discovery (unlikely, but always possible) then you may wish to claim priority for the good news immediately. One way is to write a letter to the editor of a journal which has a correspondence section, such as Nature, the Lancet and the British Medical Journal (in the U.K.). Alternatively, you may wish to contradict or confirm a statement written by someone else or to broach a subject which cannot be put into words in any other way; editors welcome controversial issues because they stimulate others to put pen to paper and may provoke lively and emotional writing, the very meat of a good journal.

Most researchers write badly. In writing a letter, which may be of vital importance to your reputation, a structure is required:

- Start with the greeting, "Sir", since all letters are addressed to the editor of the journal, even though your pointed remarks are really aimed at that chap down the road who has got it all wrong. There is no need, therefore, to send a covering letter unless you feel strongly that your reasons for writing should be made known, but not published.
- The introduction, of one sentence, should explain why you are writing. If it refers to a paper published earlier in the same journal the reference should be given, in brackets (the editor may delete this, but never mind, he will realise that you are being precise). If you are introducing a new subject, perhaps your own gem, then cite the necessary authors in the letter and provide full references at the end.
- The main message that you wish to convey should be a simple statement in one sentence, developed sufficiently to convince others or contradict them in later sentences. Simplicity is the most sought-after quality in art, so why not in writing?
- State the conclusions briefly but firmly. Most journals will

print tables, graphs and line drawings as easily as words (but do allow for the required space), so that data can be published in a very acceptable form. For new work there is no need to go into great detail to establish priority, but remember that the complete paper must appear within reasonable time, in the journal of choice, and it is then permissible to write a second letter referring to it. One editor I know says simply, "Astonish me and I will print it".

- In writing a letter you have to write personally for someone, so why not for the editor. Try to think of him and forget the 100 or so who will read your message; is he tall, short, thin, fat? Write to him as a real person, even though you have never met him. Put your case squarely to him – the preliminary results of the work, in simple terms so that he will understand. Give a reference or two that he can look up easily. In other words, concentrate on one man when writing and completely forget all the others.
- Finish by signing your name, giving your qualifications, position and address. Editors do not like printing anonymous letters, but will do so if the author in a covering letter provides serious reasons for not wishing his name to become known.

Always have something worthwhile to say and say it in simple terms. It will then get published quickly. But, for heaven's sake, have it typed and be accurate in what you write; inaccuracies are the life blood of letter writers to journals and to be crucified in public is demeaning and depressing.

LIBRARY

The man who starts a research project must consult a library, if only to discover whether his ideas have been developed before. To neglect to do so is courting disaster; the work may have been done before and perhaps later discredited, so, starting work from ignorance he may find that all he obtains are ashes. On the other hand, the previous research may have been done badly, with wrong conclusions, and if so his own efforts can be directed more accurately, more pointedly, towards the truth.

This is not universally a popular view, for some hold that, by reading publications on the subject, imaginative thought becomes biased or stultified. I disagree. The more complex the research, the more important it becomes to have access to a good library, haunt it and never to think of one without the other.

Learn to read quickly and to scan journals which may contain important information for your project. It is not easy, but time must be set aside for intellectual nourishment. The real problem is how to cover the numerous journals published every year which contain an item of value. When the research is underway the field of search is narrowed down to a specific title – which can be sought from the Index Medicus or Current Contents.

But this is not keeping ahead. The Royal Postgraduate Medical School Library subscribes to about 700 journals a year. Each week it issues a Library Bulletin of current journals (about 60) setting out the titles of every paper from each, say, 1,000 titles. It takes about half-an-hour to read quickly through the list for publications which might personally be important for the work in hand, perhaps 6–12 items per issue and one can then spend an hour in the library looking up the originals. This is one way to keep up-to-date but it does require a lot of effort at regular intervals unless you wish to be completely snowed under by all those references which *might* be vital.

Even so, one has to read discriminately. It is neither desirable nor practical to read and quote every paper written previously on your research subject. Often the most recent and the most informed will suffice.

CONSULTING THE LIBRARY

The value of a library cannot be estimated in terms of money or time. A good library is invaluable to anyone who wishes to find new material, to check old ideas, and to seek intellectual stimulation which your colleagues cannot provide. Make it your spiritual home.

If there is a medical library in your institution or nearby and you do not know it well, take time to browse around. Find out how books and periodicals are arranged and try to evaluate the library's merits and deficits; there will be both. Consult the expert, the librarian, by stating what you want to do and allow some of your enthusiasm for the subject to brush off on to her. If she has been helpful in finding references you can say "thank you" very simply by acknowledging this at the end of your paper; send her two reprints when it is published, one for herself and the other for submission with any future application for promotion. This is no more than good manners and is generally appreciated.

One should talk to colleagues working in the same field for help in finding a lead into the relevant literature. If a drug is involved, pharmaceutical companies will usually provide a wealth of references and reprints. It is also necessary to be familiar with essential sources of information: Textbooks, Encyclopaedias, Dictionaries, Review Journals, Monographs, Index Medicus, Year Books and the Medical Annual, all of which are considered in another section.

The author who is interested in making the most use of his library should read and digest *How to use a library* by L. T. Morton (1968). When asking the librarian's help, remember to:

- Allow enough time for her to obtain the text you want, if it has to be obtained on inter-library loan.
- Verify everything. Don't accept a reference from a text book as accurate; there is more than a 10% chance that it will be wrong.
- When looking for references, sharpen the question you ask by

deciding on the years of publication and in what language you wish to search.

- Discriminate in what you read. There is no need to read every article on the subject; it is equally important to know when to stop. Reading does not make a man wise; it only makes him learned.

RECORDING REFERENCES

Research is a multidisciplined and sophisticated act if you wish to do it well. You should therefore cultivate the habit of reading journals regularly and recording immediately the essential findings on cards.

- The method by which this is done is not explicit, but 20 × 13 cm. cards should be carried at all times. They should be ruled on one side only, for the essential reference, and plain at the back for notes on detail, diagrams, and tables.
- The lay-out should always be the same and in general the Harvard system is the best. Leave the top line of the card blank for key-words to be inserted later, depending on how you wish to file it. The author's names and initials should be written on the second line on the left side, and on the right the year of publication. On the third line in the middle of the card, record the title, the journal, volume number, with first and last pages. Add the month of issue for easy referral later. Beneath this write correctly the exact title.

  At the bottom left make your own assessment of the value of the paper as "good", "bad", or "non-informative" and a note on the type of illustrations and whether the list of references given are key ones for future use.
- Each card may now be filed in a box and arranged in alphabetical order. On the lid of the box one can record the title of the subject, such as "Bed Sores". The dividing cards can be labelled sequentially as: A. "Bed Sores – causes"; B. "Treatment"; C. "Prevention", and so on. Many of the reference cards will cover more than one aspect of your personal subdivisions and will have to be filed in the most appropriate section. However, since all cards are filed in one box it is probably less labour to look through other sections than to make out cross-index cards.

● When you have written a paper for publication, the list of references will be easy to compose. Just pull out each card as required, arrange them in alphabetical order and write one new card only as an example to the typist of the style in which the journal prefers to list them. There is still no uniformity in journals, and until there is the Harvard is the simplest system for personal note taking.
● Cards should also be made out for your own papers as they are published. If they are filed in a separate section, here is the basis for the annual report of your unit, with little effort.
● If you always carry a dozen cards in your pocket, you are constantly equipped to record information from reading, as well as noting interesting observations. These cards should also record reprints, photocopies and articles torn out from journals, with a note of where these can be found in your office. If you have a box-file marked "Administration" a paper on surgical technique might be usefully filed there because it also deals with the organisation in the operating theatre, a subject with which you are concerned.
● One evening, once a month, looking through your cards is time well spent, for thinking out new ideas and for rearrangement of your collection. You may decide that you have to cover the literature more widely and the only way to do this is by consulting a central agency, such as the Medlars Service.

A COMPUTER SEARCH

This is commonly called a Medlars Search (Medical literature analysis and retrieval system) and details can be obtained from the "U.K. Medlars Service, National Lending Library, Boston Spa, Yorkshire, LS23 7BQ". A handbook for users is also available. The titles of about 2,800 journals (or 165,000 articles per year) are held on tape and the computer will select and print out those you require. However there are snags, and it is too easy to be burdened with many hundreds of titles for which you have no use.

● The decision to use Medlars must be made after examining all journals available, and after consulting the librarian.
● The search must be strictly confined – by years of publication, language and the correct keyword (see Medical Subjects

index, Me SH). If you wish to consult papers on bed sores (an enormous subject in itself) you will probably not benefit from a paper on wounds, written entirely in Japanese. Key-words can often be located by noting the headings and subheadings used in the Index Medicus. Recent papers by Thornton (1971) and Barber, Barraclough and Gray (1972) should also be consulted.

LITTLE BLACK BOOK

Ideas for research come from three sources:

- Simple observation of everyday events, by far the most important;
- From reading books and journals and asking "Why?", when a statement is made;
- Contrived events – "I wonder what would happen if. . . . ?"

A permanent record has to be kept of these and for those sudden flashes of intuitive ideas which come to most of us at awkward moments. The only practical way is to carry in your pocket a little black book – soft durable cover, sewn pages and not much larger than about 11 × 7 cm.

Note down ideas when they occur, one to a page. Hence a 200 page notebook will probably last 2–3 years. From time to time, look through what you have written, add more words to develop the central theme, comment or cross out as being impractical, impossible, or now less interesting. Recently, I discovered my own 1953 edition and was pleasantly surprised to find how many ideas had come to fruition. The mere recording of thought provided a fertile soil for new and better ideas.

You may feel with Cecil Rhodes "So little done, so much to do", but without a record, how will you know? It may be necessary to list separately the "top twenty" of the most promising enterprises, and then get on with only one: it is often more important to get started than to finish.

LOGIC

If you want to read about logic in research then read W. S. Jevons on *The principles of science, a treatise on logic and the scientific method* (Dover. New York) all 769 pages. There is much of interest but little of practical use. It's a scholarly work, originally published in 1873, nonetheless out of date in its thinking if only because it tries so hard to bring classical logic into scientific method (that has evolved and is still evolving) to find answers to questions that can be put in scientific terms; it reads like a dictionary of what might have been and it fails lamentably.

MANAGEMENT

This subject is an unwelcome and unwanted aspect of work in most laboratories. However, without good management nothing goes well (and see PEOPLE). Indeed, the growth of applied technology for day-to-day problems in Western nations has been due very largely to improved methods of management: as Drucker wrote, "there are no underdeveloped countries, only under-managed ones". So too in research.

There's an old adage that "if you can't join them, beat them" and it applies here. In order to communicate correctly with the administration of an institute it's worth the trouble to read books on management (even if the opening sentence is "The new discovery for management is the use of the scientific method", which should be mother's milk to you anway). Here are three for a start:

- *Technology, management and society* by Peter F. Drucker. This collection of essays, dating back to 1958, was published by Heinemann in 1970 and is now available from Pan Books, London (1972). It is full of good sense and chapter eleven is pertinent to medicine.
- *The human side of enterprise* by Douglas McGregor (1960, McGraw Hill, New York) shows what can be accomplished on the personal level in management.
- *Up the organisation* by Robert Townsend, with the subtitle, "How to stop the company stifling people and strangling profits" is a light-hearted debunking of conventional administration, written by an astute business man who was successful in spite of other people; amusing and full of truth it's worth a railway journey just to read it. Reorganising, he says, "should be undergone about as often as major surgery. And should be as well planned and as swiftly executed". Would you not agree?

Fortified with the wisdom from these three books, talking to an administrator should present no problem; he may confuse you with the latest management jargon but you will be one ahead (if in doubt, consult the jargon chart for a suitable phrase to confound him).

MEDIOCRACY

A mediocre piece of research is of little lasting value, so set your sights high and aim only for the best. Having said that, it must be admitted that much research is mediocre. Unfortunately, it is not always possible to judge the value of a piece of work at the time, for much depends on future events and how things will change.

No-one should condone or encourage bad work nor poor composition in reporting; but the latter is remediable, the former not. Both accept mediocracy as a gaol. Mediocracy infers equality and pacifism, but militancy pays off in research – as it often does in the political scene.

Everyone must be judged on his performance – fairness, justice, call it what you like (and many institutes do not have it) – not on his looks, personality, who he knows or to whom he is related. Such judgement is the basis of a meritocracy, surely better than a mediocracy in research every time. It is important to reward outstanding performance and equally important that failures do not get rewarded, a decision which is more painful, difficult and often funked. The motto then is "Help the worthy, ignore the bad". It is soft-hearted, but usually unproductive, to encourage poor performers. This is particularly seen in the sphere of postgraduate education where many desire a higher degree, but are either incapable or unwilling to put in the effort and expertise required. The excellent will succeed in spite of difficulties, the mediocre have even-chances at best, if only because they preclude professionalism on the job.

MEETINGS

There are two kinds of meetings – the useful and the useless. Whether scientific meetings or committee meetings, the classification still applies.

Meetings are for meeting people. Some of us travel half-way across the world just for this purpose. It can become a habit, sometimes an uncomfortable habit ("Breakfast in London, lunch in New York, baggage in Brazil" – it does happen!). On the whole, if you plan to run successful, progressive, productive research, then time for travel to exotic foreign countries is strictly limited. Who will look after the lab. while you are away? International meetings are mainly a waste of time and effort, an occasion to socialise but little else. If there is some international figure you must see and talk to, then visit him on his home ground (you'll learn more that way).

There are other sorts of meetings – committee meetings. On the whole, these too should be avoided, but when you have to attend try to do your homework thoroughly beforehand. Committees make cowards of us all – simply because no single person is to blame when things go wrong, bad decisions made, or good work deferred: it is always the wish of the committee. In some situations a committee is necessary to obtain concerted action. These committees should be small, have a single clear purpose, and get through the business efficiently. If you have to set up such a working committee, see the recommendations made earlier in this book.

METHODS AND METHODOLOGY

This is the intensely practical side of research. It can only be learnt in the laboratory. So why not start there?

Very little in this book has been written about specific methods in research, partly because there are so many, partly because they cannot really be learnt from a book. Of course there are books on individual subjects with titles such as "Principles of . . . . . , Techniques of . . . . . , Methods in . . . . . , Advances in . . . .", to be read and digested. But, on the whole, methods are eminently manual and practical procedures which have to be acquired and understood by doing them.

Books describing individual techniques are helpful, but you will have to attend courses and tutorials (if you can find them conveniently nearby), go to watch others at work, then join in or try on your own to repeat what you have seen. The best way is to apprentice yourself to a master to learn the craft of manipulation.

In recording the methods used, one has to state specifically:

- The details of apparatus, materials and equipment used. Diagrams and illustrations may be included to advantage.
- The procedure followed: a step-by-step account of what was done and why.
- The observations made and how they were presented (as tables and graphs).
- The calculations necessary to give meaning to the observations.

Methods should be detailed enough to allow another to repeat the work without having to search unduly for specific procedures; if the latter are well known and generally accepted, the relevant reference will suffice. If, however, any method has been modified such alterations must be described in detail, usually with an explanation of why; not all modifications are better than the original.

MISCONCEPTIONS

There are several misconceptions about research. I heard them 25 years ago; they are still fresh today and just as far from the truth.

- Research is easy.
- It comes naturally to those doing it.
- It requires no great intellect (or genius, depending on the speaker) and no particular skill. Hence, the man who is not much good in clinical medicine should go into research; it provides a quiet niche for the less bright.
- No discipline is required.
- Research is divorced from clinical practice and therefore unimportant for the clinician or the welfare of the patient.

MISTAKES

We all make mistakes. Often they are quite simple, so check on the obvious first: erudite explanations rarely apply to everyday events. Mistakes are everyone's fault, so stop looking for a culprit and start putting things right.

The astute will analyse why a mistake occurred and may even profit by it. Indeed, in medicine more advances have been made by mistakes than you might think. The team leader, director of research, or whatever he calls himself, is *the* one responsible for disasters, dissents, quarrels, mistakes. Moreover, he has to defend the delinquent, find a solution, apologise with dignity, yet make sure it does not happen again. Not an easy task, but vital to success.

NEVER

There are an awful lot of these about, from "never say always" to "never say never". Here are ten that may help in research:

- Never criticise another researcher, only his work. He may have more friends in high places than you.
- Never show anybody a half-worked idea; finish the thinking and write it out in precise language first.
- Never talk about brilliant projects until you have the work in hand, for another may beat you to the post and publish first.
- Never undertake to do work for a commercial company unless you are fairly certain of a result either way.
- Never forget to acknowledge the finance from a grant-giving body and those who helped materially (the Chief, Medical Artist, Technicians, Nurses); better still, put their names on the paper for publication.
- Never acknowledge help that was not given, even if you think it will please (it will of course, but you may bitterly regret it later).
- Never forget that you *might* be wrong, even if your name is Wright.
- Never promise to deliver work by a certain date, unless you have already completed it.
- Never pursue a lost cause. If others have been unable to solve a problem and you clearly cannot, then move on to something else, while you are still healthy.
- Never give up.

Now add your own in the space below.

NOTEBOOK DISCIPLINE

Two basic rules apply to writing in research:

1. Every observation and every experiment must be recorded in writing at the time they are made, preferably in a day-book so that precious data cannot be mislaid. Good note-book discipline is essential. When there are sufficient data, tables can be constructed and graphs drawn on squared paper stuck into the book. Tentative or preliminary conclusions, made perhaps after only a few experiments, should be recorded. Later, summaries of progress from a larger series of experiments will help to concentrate your thoughts and to be critical of your own work. If there are research-in-progress meetings at your institution, new ideas pertinent to your own projects should be written in the same book. Remember always to record the date at the top of the page whenever you write. A well documented note-book is a personal record of action, thought, and time – a precious thing. When it comes to writing a paper, the direct transfer of the results from the laboratory note-book to paper without considerable editing will not produce an acceptable publication.
2. Findings should be published as soon as a significant advance in the subject has been made, and not before. To wait until all questions have been answered and all possible hypotheses explored may take a life-time. To man no end is discoverable; everything is relative and nothing is certain. The technique of writing for publication requires a special discipline, but one would point out that much research is written up too quickly. It is stupid to believe that research which has taken two years to complete can be written satisfactorily in two weeks.

OBSERVATION

Another over-worked word. It has been mentioned before, but its importance should never be under-rated; it is the first step in any scientific enquiry and the most important. It is also the most elusive, for we see only what we know and the commonest remark of a visitor to our labs. is, "What am I supposed to look for?". A sad reflection on our forefathers and on the whole theory of the scientific method. Here are 10 points to remember:

- All observations are biased.
- Observations depend on looking and seeing – a conscious effort.
- Observations can be erroneous: optical illusions if you like, but erroneous all the same.
- Observations can be made without apparent record: they may be recalled years later, hence the anecdotal stories of retired professors.
- Observation depends on personal interests at the time. Interests can be awakened later, but only a very observant mind will record the unnecessary and extravagant at the time.
- Spontaneous observations high-light an imaginative mind. Too often this is a "butterfly" mentality and so the observations have no meaning. Hence one hears the expression, "Well, I thought of that years ago". The answer is simple: "Why did you not do anything about it?".
- Observations can be deduced (or extracted painfully) from experiments.
- Relationships are observations.
- Everyone can be trained to observe.
- Finally, look for the unexpected, the unusual, in daily life. As Louis Pasteur recorded: "Chance favours the prepared mind". It worked for him!

OFFICE HOURS

If your idea of long-term research is 36 hours, don't enter the race. A great deal of progress is made on the "stop-go" principle, and when you "go" it may mean long hours, short meals and not much sleep. Even the "stop" is not complete immobility: thinking, talking, planning, reviewing of results have still to be done – that's the time to sort out what you have done and what might be done.

ORDER OF INVESTIGATION

Many people want to do some form of research (and there are many facets to the subject) but do not know how to start. This is absurd because the precise order of procedure has been defined by the "scientific method". There are 5 stages, all clear-cut, and it is not advisable to proceed further until you are clear in your own mind at each stage. For this reason alone, much research has to be postponed or abandoned and it is better so.

- Make an observation, as accurately as you can and try to repeat it. Observation is the most difficult step because we tend to see what is expected. To see, objectively and dispassionately, is the most difficult thing in the world to do. The accuracy of an observation is not just important, it is absolutely vital; to proceed to the next stage with incorrect observations is tantamount to failure in the whole process.
- Now try to correlate the events which explain the observation (an hypothesis). Usually there are several, but if only one then look for others because you are going to have to test each in turn.
- Experiment and comparison. These go together.
- Induction. Now you have to sit down and think. Does my hypothesis stand up? What have I learned? Does it all fit with contemporary knowledge? If it's all quite new (and you may be lucky) the planning of new experiments to substantiate your ideas becomes urgent, so get on with them now.
- Conclusion. If all has gone well so far, then the conclusion will suggest a new hypothesis. Now you really are involved in research – and just beginning, by realising that you have hardly begun to understand the vast range of marvels which the world contains. In fact, every new discovery seems only to open up a half-a-dozen new mysteries.

Now is the time to talk to people (and to keep on talking for others have made observations too) to crystallise your ideas.

ORGANISATION

The importance of organisation in research cannot be overstressed; lack of it will certainly undermine your efforts, waste your time and guarantee non-productive work. Research is essentially organised effort. So the sooner someone devotes time to organisation the better; the alternative is disorganisation, or at best the non-result and nobody wants that. There are some central principles in organisation (see "People") and certain practical points:

- The organiser must be an active member of the research team. Even if he has to sit behind a desk to think and write the script for the project, his job should essentially be in the laboratory. How else can he tell that his plan is practical, effective or even possible?
- The organiser must know how and when to communicate his ideas, at formal meetings, in the laboratory, during an experiment, or in the bar when everyone is relaxed but probably tired at the end of the day.
- No organiser shall be entirely deaf to modification of his plans. He should ask each worker what he can contribute.
- The organiser should be concerned mainly with results, not effort.
- The organiser is accountable for the whole to his superior, whether this is the director of a department or the secretary of a grant-giving body.

The organisation of a modern hospital or research institute is a vast complex of interdependent relationships, up, down, across, and even diagonally. In practice, the interdependence is so great that only collaborative team effort can make the system work effectively. Hence, when organisational charts are drawn up (a favourite past-time of professional administrators) relationships should be shown as linked groups and not a hierarchial structure of one individual reporting to the man above.

Good organisation is largely a matter of producing good managers, a matter of making them "grow" in a suitable environment because they cannot be manufactured like material products. People who try to mould a junior to their own way of thinking and doing, only produce a mediocre facsimile. The "agricultural" approach (of keeping the soil in good tilth and suppressing the weeds) is the only worthwhile method to develop. After all, the chief purpose of encouraging people to do research is to increase their ability to learn from experience. For this purpose, statistics and records of performance are only of value in keeping track of performance and not simply as a means of developing talent. Training occurs only on the job, but must be rewarded by time off, merit money or salary advancement (going up the queue) as bonuses for work well done.

One person with only half a job can wander around and do real damage in his or her spare time (the Salmon No. 7 in nursing is a good example where embryo administrator nurses wander around with notebooks looking for complaints; they are trouble-finders, not trouble-shooters). The best organisations are slightly under-staffed (for instance, the autonomous out-patient surgery unit at Hammersmith Hospital is run by four nurses yet manages to carry out one-fifth of all the surgery in the place, plus removal of sutures, entertaining occasional visitors and the secretarial administration, yet demands more work). It illustrates the proverb that if you want something done quickly, then ask the busiest man in the outfit. In the same way a tight budget brings out the meanest and the best qualifications in a person, because he will simplify the organisation when under pressure (see Townsend 1971).

OVERWORK

Your mother, wife, girl/boyfriend will accuse you of overworking. Maybe. If it is true then you will have to change your way of life. There are 6 ways of lightening the load:

- Organise, to off-load to another, all those things which do not demand your attention. Routine mail can be dealt with by your secretary and she can take all 'phone calls with a promise to call back within a certain time – expensive perhaps, but so is your health.
- Delegate as much as you can.
- Insulate yourself when you have to concentrate.

At the Royal Postgraduate Medical School there are carousels beneath the main library, accessible only by a spiral staircase – no telephones, no noise, only musty old volumes for company. Loneliness may attack you, but there is the chance to think alone.

- Concentration is the most difficult of all the arts. How to do it? Good advice is to complete the creative work first, before getting down to routine. Those who travel fair distances to work will probably have started soon after leaving home, so the office is the place to continue the same train of thought.
- Mechanisation when possible will relieve the daily grind. A tape recorder for your secretary is essential: have two if necessary, one for routine letters, the other for notes on research. Both purposes can be, of course, accomplished by a good supply of cassettes for one machine.
- Eliminate everything of little importance. A large waste-paper basket is essential equipment for all research. You will have to decide not only what can be thrown away, but which committees to miss, which social occasion to avoid, which person to dodge. It requires courage and tact, so pass the job to your secretary. Your job is to do research.

OTHER ACTIVITIES

Researchers tend to be a narrow-minded lot. Research occupies most of one's working and leisure time (certainly in thinking about it), so it is all too easy to become separated from the real world around us.

It is important to have a hobby and to talk about it; important to take interesting exercise and talk about it; important to take an interest in the arts (music, painting, sculpture, literature) and to talk about them. To discuss politics, people, economics and to have distinct views on each, is as important as an interest in your own research project. The mind is broadened away from the narrow outlook, the body kept fit and the spirit enlightened.

PARAMETER

Never use the term when you mean "variable", which is more often than not the case.

Never use it if it can be avoided however popular it may be; parameter gets up some people's noses, like house dust, and evokes emotional correspondence (the word is not indexed in Bradford Hill's *Principles of Medical Statistics*, 9th Edition, 1971, but it is in the *O.E.D.*)

PEACE KEEPING

Every team worth its salt will have quarrels. That is only human nature, and often human nature under stress; unrecognised or untreated they quickly get out of hand and become almost uncorrectable. Some thrive on conflict; watch them.

Most ill-feelings arise in three particular areas:

- Lack of credit for originality. This should be overcome quite easily by publishing a short paper, with the authors in correct order (and usually that is *not alphabetical*) to establish priority. If an invention is marketable and patentable (two quite different things!) then establishing the originator is important. So often, of course, an idea gets thrown around, modified, tried out, so that at the end of the day there is no clear notion of who really was responsible. In these circumstances it is better to accept a new invention as a team project and to avoid linking anyone's name specifically to it.

The trouble comes when the product is really new and could be patented. Money is involved as well as prestige: this is deep water. Divisive elements enter quickly but the position has to be faced. If there is trust and faith in each other, solutions are rarely difficult.

- Inequality of effort. When one is working hard, fully employed during the day, with another nearby who spends much of the time reading a novel and clearly under-used, frictions are bound to develop. This situation is best anticipated and prevented. It is the duty of the director of the research team to seek it out and apply corrective measures. One way is to divert the drone to another area, or dismiss before serious damange to morale has occurred. Empires may be built by employing a large labour force, but they are also lost that way, from sheer discontent.

In developing a research team it is better to start small and only employ extra staff when it is abundantly clear that there is a full day's work to be done by another pair of hands and for a reasonable period of time. For instance, in the out-patient surgery unit at Hammersmith, run by only 3–4 nurses and an orderly, one-fifth of all the hospital surgery is performed, compared with a staff of 36 for the remaining four-fifths. All are fully occupied yet a secretary on site would be useful, but not fully employed. Hence there is still no secretary, simply because the disruptive effect of someone with time on her hands could be so disastrous. There are occasions, of course, when research does not go with a swing. It is then better to give days off, in rotation, as bonus for work anticipated.

● Privileges: time off for extra hours spent on research is accepted generally. The trouble comes when a day off is a reward for special effort. It is vitally important not to reward slackness, yet to make it clear that outstanding effort carries its own compensation. This is the difficulty. In most institutions it is not possible to increase the salary of those who work harder: convention is against it. However, there are other means. Sending a person to a meeting (with all expenses paid) is one way. The ideal, of course, is that all will contribute equally in time, effort, and interest – a myth if ever there was one. When it happens, then why not an evening out on the house (at the Director's expense)?

Who does what, a demarcation dispute, has been an issue among Trade Unions for years: almost a national joke. But who tells who what to do is quite a different matter: here we are involved in competence, status, delegation. In a research team it may be difficult to tell them apart; in a good team it rarely matters. The shifting leadership principle takes care of most situations. In the final court of arbitration the director of the research team may have to adjudicate. He should never let the sun go down on such issues – they are urgent and must be solved immediately. There is no other course.

PEOPLE

All medical, paramedical and biological research involves people – colleagues, team-mates, patients and relatives. It is therefore essential to know how to handle people, to know the principles of good management. They are easy to understand, the people and the principles, if only one tries. Both are often overlooked or ignored.

The word, the big word, for this simple subject is "ergonomics" – a term coined for industry where the management of people can make or break a company. Doctors, on the whole, manage well because they have compassion, knowledge and a realisation of biological inexactitudes. They become poor or bad managers when they fail to recognise that there *are* principles of management which have been evolved and have become established since the technological revolution. The man who continually talks about "principles" probably possesses none, but in research it is important to sift out those who are productive and those who are not; eliminate the drones if only to preserve the morale of the workers. The problem is how to deal with people.

In industry, over the years, 10 principles have crystallised to make life easier and results more certain, to increase productive efficiency. Some of the principles have been recorded in a light-hearted manner but this does not mean that they are trivial.

● *Pharoah's Law* states that slaves ask two questions:
 1. Do they keep us busy?
 2. Do they know what they (and us) are doing?

If the answer to both is "yes", the job gets done well (a pyramid is built); if the answer to either is "no", the job is botched up somehow, which may not be spotted immediately. It has happened in industry deliberately and I suspect in research too.

● *Parkinson's Law* declares that work increases to fill the time available for its completion. Hence a man will spend a whole day carrying out one test, even though he could easily have done five; the fact that there are not five tests to do is usually the fault of the organiser. It is often wiser to say "Go home" rather than allow time to be wasted in this way.

● The *Peter principle* points out that in a hierarchy everyone is promoted to his level of incompetence (The cream rises to the top until it sours). There are precedents in every institution and it would be invidious to quote individual examples. Just look around you!

● *Townsend's time sharing*, or the 80/20 principle, records that in daily work 80% is routine and could be delegated, whereas 20% requires most of one's time, thought and energy – as true for the researcher as for the Company Director.

● *McGregor's theorems* assert that in management there are theory X and theory Y people who hold the following convictions (or beliefs, because religious faith and tradition seem to be firmly incorporated).

Theory X is the traditional view of direction and control. It specifies that the average human being has an inherent dislike of work and will avoid it if he can; that most people must be coerced, controlled, directed and threatened with punishment to get them to put forth adequate effort towards the achievement of organisational objectives; that the average man prefers to be directed, wishes to avoid responsibility, has relatively little ambition and wants security above all.

Theory Y represents the integration of individual and organisational goals. It assumes that the expenditure of physical and mental effort in work is as natural as play or rest; that external control and the threat of punishment are not the only means for reaching a target. Man will exercise self-direction and self-control in the service of objectives to which he is committed; such commitment is a function of the rewards associated with achievement. The average human being learns, under proper conditions, not only to accept but to seek responsibility. In other words, he grows in ability in the right environment.

Douglas McGregor (Sloan Professor of Management at M.I.T., Boston) goes further and believes that the capacity to exercise a relatively high degree of imagination, ingenuity and creativity in the solution of any problem is widely, not narrowly, distributed in the population. Under the conditions of modern life, the intellectual potentialities of the average human being are only partially used and rarely discovered.

- *The Manipulator's Law* covers various measures that can be, and are, taken to change the structure, behaviour and environment in a research team. Certain euphemisms are commonly heard, such as "gaining experience" (which means moving people from place to place frequently), or, "the best in the world" (because there are no comparisons, nor real knowledge; this is a statement of utter blindness to faults and much needed change). They are a form of self-deception, a variant on Stephen Potter's (1935) "Gamesmanship", but it does allow a dud director to stay at the top long after he should have gone.
- The *Surgical principle* asserts that cancer starts small, spreads outwards and becomes incurable unless measures are taken early to confine it. This applies to morale in any area of personal activity whether there is an end-product or not; if there is an end-product the results are noted sooner. Morale, like clinical acumen, defies quantification, but it cannot be ignored and has to be nurtured.
- *Drucker's Law* proposes that the distance between the leader and the average is constant, hence the greater the leader the higher the average performance. This is why we see centres of excellence (and mediocre ones too) in every field of endeavour.
- *The Law of Elasticity* affirms that unless people are occasionally pushed to the limit, no-one really knows what they are capable of. We do know that most ordinary folk are capable of extraordinary feats of courage, endurance and quick thinking in times of war and crises. If extraordinary achievements are wanted from a research team then stretch them, but remember to relax as soon as practicable afterwards.
- *The Law of Expectations* maintains that people's expectations (and not what actually happens or what they get) lead to happiness or frustration. Hence, promises not only have to be kept, they have to be realistic to start with.

If the researcher works in an institution, as most of us do, then he has to make time to get to know everybody by name who works there, from the highest to the lowest; they should be greeted when met, the former require "Sir", the latter "Fred" or

"Jack". There is only one time to do this, when you first join, for later may be too late and be misinterpreted. It is, however, a good investment of time because quite often research may demand the help of a carpenter, plumber or electrician and it's usually urgent. People respond more readily to those they know.

PERSONNEL

As soon as you start talking about those who work with you (and some work for you, poor souls) as "personnel" you have acquired the beauracratic attitude. Now is the time to take a good purge, retire for a time ("temporary illness in the family" will do) and rethink your own position. If you are sitting on top of a dung heap and are proud of it, you are a mushroom and not a leader so move over. If you are sitting beneath a dung heap (that is, others are telling you what to do), take a long holiday, watch the birds and the bees and reconsider whether you are in the right institute. Tough, isn't it? But it has to be.

PILOT RUN

Before embarking on any research project, clinical or laboratory, the wise man will try his scheme out in miniature. From this it should be possible to find out and forecast what may not have been obvious during the planning stage.

- Apparatus: Who will work it? Is it reliable, fairly robust, portable? Are measurements accurate and reproducible? Electronic equipment has a habit of going wrong unexpectedly so who can repair/correct it?
- Data: How long does it take to handle? Is information required immediately? Can it be stored? Is a computer required to sort it all out?
- Programme: What are the likely/unlikely snags? (they must be looked for). How much time does each procedure take? How long to fill up forms? Is the project really too complex or too difficult for the resources available?
- Cost: Is the project going to be more costly than originally thought?
- Completion: Is the planned date for completion likely to be met? Will finances be sufficient?

If difficulties are discovered early on, then all may not be lost because the project plan can be modified or completely revised. The young man, eager to get on, may not readily accept the idea of a preliminary experimental project but it can be an insurance against later disappointments.

PLANNING RESEARCH

There are three forms of planning in research:

- Tactics, that is, the actual conduct of an investigation and usually best left to the investigator.
- Strategy or planning further ahead. The whole research team should be involved so that each may comment and advise. Centralisation of a series of investigations, with independence for the individual, has advantages.
- Policy is often best decided by a committee, a small committee, who can plan the problem to be investigated from available funds, and realise what further monies will have to be acquired.

All three types of planning should be stated in writing so that everyone is clear and no-one has to rely on memory for details. Unfortunately, although research can be planned fairly effectively, discovery cannot.

The scientific investigator must have some idea, essentially a philosophical one, of how he is going to set about acquiring an understanding of nature before he can apply himself to the task. In practice, he may be entirely disinterested in philosophy, preferring to regard himself as a compiler of demonstrable facts. He cannot, however, escape the implications of adopting a definite scientific method, which teaches him to record particular kinds of facts, by using certain recognised procedures.

We are concerned here with the overall plan, including its administration, of how the scientific method should be applied. All medical investigations can be divided roughly into three parts:

- The planning of the experiment.
- Its execution and the collection of facts.
- The interpretation of results.

Of these, by far the most important is the stage of planning. At this time the whole future of the investigation is decided; with a good plan all may go well; with a bad plan or an ill-considered one, the work may never be completed and the opportunity to repeat it may be lost for ever. Time in setting up a good project is well spent, rather than trying to patch up a bad one later.

The plan of an investigation is the protocol (the code of practice, if you like) by which the investigator decides in advance how the investigation shall proceed, in what order and what observations shall be made. There is rarely success without a good plan. It is often helpful to have a set questionnaire to be completed at an early stage (see PROJECT PLANNING FORM). This outline plan may occupy only a couple of pages, but it will provide a nucleus to which further detail can be added. Research has its own momentum – when things go well, they go fast. Hence, the demand to think out detail, draw up a protocol, even carry out a pilot experiment, before the main work is undertaken. It should also be noted that postponement of a project is usually a lethal act; the resurrection of old plans is rarely worthwhile.

Planning, then, decides in advance:

- the what, when, where, who and how of a project,
- its objectives, which at first should be limited in extent and eminently obtainable,
- the policy of the course of action necessary
- the procedures, that is the methods, to be adopted
- the programme which outlines the steps in an orderly manner to accomplish the mission in a specified time
- and finally how the expenditure of money, labour and time will be spent (the budget).

PLEADING YOUR CASE

Some charitable bodies have a scientific committee who not only vet applications but ask applicants to appear in person before them. This can be quite a daunting interview. It is important to bring a copy of the application and to have re-read it carefully beforehand so that any questions of detail may be answered with confidence. If, however, some work has already been carried out to support your ideas, the results can usefully be put across to the committee.

Large white cards with the information written boldly with a broad felt-tipped pen may effectively demonstrate what you have found; these need not be done by a professional artist, indeed, the mere fact that you have bothered to bring them in rough should show that you understand the difficult decisions that the members have to make. After all, an architect attending a similar meeting would naturally bring his drawings for display. It is important to resist the temptation to turn the data into lantern slides for you will then have to bring a projector, fuss about finding a suitable electricity supply and may have difficulty in finding wall space, uncluttered by pictures of previous benefactors, on which to display your wares.

Alternatively, write the tables of results and the main headings boldly on a roll of plain white wallpaper (lining paper is cheap and satisfactory) but remember to bring a rolled umbrella with a curved handle to thread through the roll, which can often be placed on a convenient window-sill (or even held in the hand) as the story is unfolded. If the research application concerns a piece of portable equipment, then bring it with you to hand around for all to examine.

When the interviewing body has several candidates to see (look around for other edgy people while waiting your turn), brevity is appreciated; hence you need not say very much. Certainly, this is not the occasion on which to start a spirited lecture on the importance of your own work. To hit the right note requires some thought before the interview.

PROJECT PLANNING FORM

To anyone running an active research unit, even if it is only one man and a boy, others will apply to you for help; either they want advice on how to start or the offer of accommodation. For such applicants it is useful to have a simple form, which they can take away and fill up, so that you will have a clear idea of what it's all about and can then discuss the proposition in detail. Many people cannot express their ideas clearly in words and may find that they are impractical, anyway, when written on paper. Suitable headings (with appropriate spaces between them) can be placed on two sheets of A4 (297 x 210 mm) paper, which not only makes the applicant condense his thoughts, but saves the receiver from having to wade through pages of verbose prose.

- Name and date
- Proposed work/experiments
- What I expect to find
- Clinical significance or application of findings
- Work previously done on the subject by others (2–4 references)
- Criticism of the work done by others
- Why I wish to do this particular project
- Material to be used and quantity
- Methods of the experiment
  1. Experimental design
  2. Experimental technique
- Duration of the experiment
- Special technical assistance required.

PROOF

Sir MacFarlane Burnet wrote: "No such thing as absolute proof exists, but there is always contemporary truth in the form of axioms, generalisations, theories and hypotheses that, in the opinion of competent scholars, have not yet been disproved." (In *Reflections on Research and The Future of Medicine* Editor: C. E. Light, 1967 McGraw-Hill, New York). So why talk about proof? The answer is simple, "Don't". Far better to talk about probabilities, worked out from the evidence supplied.

Proof and perfection exist only in theory, not in practice. We have, therefore, to accept a qualified perfection somewhere within the tolerances permitted by the specification of the work to be carried out – a permitted variability within certain limits. The only way to do so is to use statistics sensibly.

PUBLICATION

When is the right time to publish your work? When all the answers are known? No, for this may take a life-time and, anyway, most answers generate a new series of questions, so the line has to be drawn somewhere finite. The correct moment to publish the findings of an investigation is as soon as a significant advance in the subject has been made and not before. Even so, it is worth starting to write before all experiments are completed if only to retain the chance to do further work needed to support what you state.

Much research is written up too quickly. It is stupid to believe that work which has taken two years to complete can be written satisfactorily in two weeks. When the time comes to write a paper, good organisation is important if the job is to be carried through in reasonable time and with the least irritation. If the communication is preliminary, a suitable structure would be as follows:

- Introduction
- Methods of study
- Results
- Discussion and conclusions
- References

So start with 5 separate sheets of paper, each with one of these headings and then jot down all ideas that are appropriate.

When the publication reports the full results of work done, the following stereotyped structure must be followed. Some people find it unduly restrictive to good writing but it does allow the reader to find his way easily in all publications, it follows a logical order (even though the work may not have been done in that order) so that the reader can understand the argument and agree the conclusions and it helps the writer to discover if anything has been left out (or too much put in, by way of repetition). See WRITING to know how to go about the job, but here is the usual format with comments on each section.

- Introduction:
Hold the reader's attention – how, why, what, when, where, and who?
- Material:
who, what, why.
- Methods:
experimental technique,
experimental design,
methods of assay. Protocol of assessment.
- Results:
separate fact from opinion.
- Discussion:
of work presented – introduce no new facts
of work related to that of others
of its relation to other concepts
- Conclusions:
1. It confirms what everyone knows
2. It confirms what people have suspected
3. It's new and has never been known before.
- Summary/abstract:
this is what is read and extracted by review journals.
- Illustrations, tables and legends.
- References to the literature cited. Use the Harvard system – name, initials, year of publication, journal, volume, page; unless the journal specifies otherwise.
- Acknowledgements:
for the diagrams, photography, technical and financial help.

Some people believe that publications from a research unit follow the law of the energy generated by a motor car when hitting a brick wall (Energy = $\frac{1}{2}$ mass × Velocity ^{2}). Hence, short papers are produced at rapid intervals. Although speed of publication is often important, to establish priority, a light-weight paper will carry little conviction and is to be deprecated.

PUTTING OVER IDEAS

There are occasions when complete reorganisation of work, or a project, has to be done. The problem is how to carry others so that the new idea is accepted. There are 10 golden rules:

- Don't assume that people want ideas just because they say they do. In most people this is purely a primitive defence mechanism for those who cannot or will not think for themselves.
- Don't believe that other people will think the way you do – they don't.
- Decide whether you want your ideas accepted or whether you merely want credit for them. The latter is often called "leadership from behind", a dangerous concept because leaders should, by definition, be out in front.
- Arrange for someone else to present your idea: you will hear unexpected and often helpful criticism which otherwise might have been submerged.
- Be casual in your approach.
- Don't confuse people by presenting more than one idea at a time.
- Use decoy ideas and throw them out when they have served their purpose, in order to have your own idea presented in a good light.
- Don't overstress the originality of your idea – the more original it is, the more important it becomes to sublimate it. No-one likes change and no-one likes you to make it obviously necessary; thus find a precedent in history, the older and the more improbable the better, for no-one likes to be out-of-date either.
- Make your idea water-tight by providing sound evidence – which does not make it unsinkable, but merely raises its credibility.
- Try a warm emotional appeal by mentioning other research units who might get there first. Alternatively, ask some disliked

person to oppose it (but now we are encroaching on "Gamesmanship" – which see).

So far, most of these rules have been rather negative and criticism of new ideas is mainly that – unproductive at best – which is incorrect because helpful hints are what are wanted. Benjamin Disraeli, a great politician in his time, once said, "It is much easier to be critical than correct". How true!

RADIOACTIVE ISOTOPES

Isotope is the name given to a chemical element which has the same atomic number as another element and occupies the same position in the Periodic Table, but whose atomic weight differs from that of the other element (by a whole number because of a difference in the number of neutrons in the nucleus). For instance, Copper has the atomic number of 29 but two naturally occurring isotopes are found (63 and 65), while Titanium with an atomic number of 22 has five isotopes (46, 47, 48, 49, 50). Isotopes are identical in their chemical properties.

Radioactive isotopes have a nucleus which is unstable, emitting α or β particles or γ-rays while changing into a more stable form ("degradation"). Few occur naturally in nature such as radium and uranium, so that most are created artificially in an atomic pile. These vary in their half-life (that is, the probability of decay within unit time) from seconds to thousands of years.

Radioisotopes have a valuable place, in research and clinical medicine, for diagnosis and treatment because the radioactivity emitted may be measured with fair accuracy using an external counter or a well-counter (for fluid samples). Moreover, isotopes can be conveniently used to label substances occurring naturally in the body, such as albumin and fibrinogen in the blood, thus making the tracing of movement and metabolism of these substances a simple matter. If the activity of a known quantity of radioactive isotope is fixed to another substance and then mixed with a known quantity of the same substance present normally, the proportion of radioactivity measured gives an accurate estimate of the total in the body. Hence, for instance the total circulating blood volume becomes relatively easy to measure.

Moreover, two different isotopes of the same chemical may be used simultaneously for it is possible to count radioactivity at different voltages. The whole subject is so important that most large institutions have a special Department of Medical Physics to deal with isotopes, counting techniques and the protection of personnel. The catalogue issued annually by the Radiochemical Centre (White Lion Road, Amersham, Bucks) provides a list of radiopharmaceuticals with costs and supply details; the centre also produces a series of monographs on the specialised uses of some of their products, free of charge. For example, Medical Monograph No 2 deals with measurement techniques in the clinical application of isotopes and No 9 with radiochemical methods in enzyme assay.

Two useful introductory books are *Principles of the use of radio-isotope tracers in clinical and research investigations* (by R. Oliver, 1971 Pergamon, Oxford) and a 100 page book published by Hospital Medicine in 1969 (250 Grays Inn Road, London) *Diagnostic uses of radio-isotopes in medicine.* It is always worth talking to a Medical Physicist and inviting his co-operation before undertaking any investigation requiring the use of isotopes, for it is possible to get into trouble just as easily and as quickly with this form of technology as it is with statistics. Everyone using isotopes or radio-chemicals of whatever form, requires a licence to do so in almost every country of the world; supervision is not just advisable, it is compulsory.

RANDOMISATION

Randomness is important because of the "fairness" and lack of bias by which a sample represents the population under inspection. More important is the fact that *the pattern of sampling variability for any population* (a group of patients) *is known if, and only if, the sampling is random.* If a sample is random the mathematical laws of probability apply and so make it possible to know the patterns of sampling variability in terms (mean and standard deviation) by which the sample must be interpreted. Random samples are the only ones which have a *known* pattern of variability. This is the root of statistics and so any data presented to a statistician which is not collected in a random manner is probably valueless. A mean and a standard deviation may look good on paper, but be nothing in practice. There is no safety in numbers! There is something about numbers that lures people into thinking uncritically; there are many examples in medicine, for, unlike armies, biased samples get you nowhere.

Randominess does not produce perfect uniformity; just the opposite, it produces chaotic irregularity, but in the long run a regularity about which we do know something. The fundamental importance of the "normal distribution" in statistics, however, arises from the fact that the measures computed from samples tend to be normally distributed whether the original data are normally distributed or not. The dispersion of a normal distribution is measured by its standard deviation (the Greek sigma, σ). Hence, if I report on a series of 100 patients and record the mean height as 180 cms $\pm$ 5 (the standard deviation) you will know that 33% were 1 standard deviation either side of the mean (175–185 cms), that 66% were within 2 standard deviations from the mean (170–190 cms) and that 98% were 3 within standard deviations from the mean (165–195 cms). Tall people indeed! They were all Police Cadets, and this is a neat way of conveying information.

Patients can be allocated at random to one or another treatments by reading from a list of published random numbers. If the total intended to be treated is 100, then the first fifty numbers read can represent treatment "A" (that is, half the total, have been recorded) and the remaining 50 places are then allocated to treatment "B". This means, however, that if the trial or experiment has to cease after only 80 patients have been seen, the totals in each group may not be identical – indeed, the last 20 patients may have been due to receive only one of these two treatments.

It is therefore wiser to divide the total into blocks of say 10 each and allocate random numbers within each block. In this way it will be known that if the work has to stop, then when the number of patients is a multiple of 10, there will be equal numbers in each group. This is known as restricted randomisation and is a very useful tool if the condition being investigated is uncommon or when the results of a new treatment are uncertain. You only have to choose five numbers from 1–10 (say 1, 3, 6, 8, 9) for treatment A, the remainder being allocated to treatment B; the process may now be repeated for the second 10 but with different allocations which are far easier to take from a list of random numbers (to be found in most books on statistics).

One of the best checks on the effectiveness of randomisation is to compare the mean height weight and age of the control and treated groups; they should be practically identical. Allocating patients at random has other advantages by:

- excluding our own personal likes and dislikes,
- excluding our lack of balanced judgement,
- removing various factors from our control (bias),
- removing the danger of trying to balance natural enthusiasm,
- removing some responsibility from the observer, so that all information gleaned is based on objective measurements.

READING

Scientific language and literary language were virtually one and the same when Quiller-Couch gave his lectures on the "Art of writing" in 1914. Since then, they have diverged in content, structure and style; the speed of divergence is increasing annually. Hence, if you only read scientific journals you will know nothing of another world which has a lot to contribute.

Read widely, for most researchers are a narrow-minded lot. It is just as important to know how a large international company is managed as it is to know how to measure the performance of your three-man research team – the latter is the microcosm seen down the wrong end of a telescope.

Read something every day: your subject, leaders in reputable daily newspapers, a book. The object is to widen your vision as well as to keep up-to-date in your chosen field. Use both the specialist library and the public library in the search for the printed word. A book has qualities which give it great force as the most free means of communication and expression of ideas, as a vehicle of culture and knowledge. Literature and scholarship would suffer without it. In spite of the technological revolution and the forebodings that books would disappear, more than ever before are being published, for the contents can be read repeatably without being changed or diminished. A book is also the most portable source of information; it can be read in a train, bus, bath, office and so is unlikely to be surpassed by radio, tape-recorder or television. The three latter may supplement reading but are no substitute for a very efficient means of storing information.

It is not easy to provide a reading list for the man who wants to learn to appreciate good prose (poetry should be included too) and the feel of words, in the same way that it is difficult to recommend music apart from the half-dozen greats, but here is a list which will start you in the right direction. It includes many short stories, simply because they are easily available in cheap editions and may lead to further reading of such authors' works. By "reading good literature" is meant reading slowly to appreciate the apt phrase, the crisp description, the epigrammatic imagination, the profound thoughts, the talent with words that marks the creative artist in literature. These books are not meant to be scanned like a newspaper and then cast aside; they are to be savoured and reread from time to time with pleasure. All too often rereading reveals hidden depths, missed on the first occasion and now discovered with excitement. They are not in any order of merit but rather as they come to mind:

- Guy de Maupassant: *Selected Short Stories*
- Anton Chekhov (a doctor of medicine): *Lady with the Lapdog* and other stories (includes a beautiful description of a medical lecture too).
- Somerset Maugham (another qualified doctor): volumes of collected short stories.
- O. Henri: *The Gift of the Magi.*
- D. H. Lawrence: Selected essays.
- L. Tolstoy: Short stories and (surely) *War and Peace.*
- Rudyard Kipling: definitive edition of poems.
- Gibbon: *Decline and Fall of the Roman Empire* (tends to send people to sleep but it does contain fine descriptive prose).
- Francis Bacon: Essays.
- Palgrave: *Golden Treasury*, poems again because words matter.
- Gerald Manley Hopkins: poems, edited by W. H. Gardner
- Ernest Hemingway: *For Whom the Bell Tolls.*
- Muriel Spark: *The Prime of Miss Jean Brodie*
- Flaubert: *Three Tales*: translated by R. Baldick
- John Betjeman's poems, so modern and descriptive.
- Alexander Solzhenitsyn *Stories and Prose Poems*: translated by Michael Glenny.

(Two sentences . . . "some with immature, moronic expressions of totally unfounded self-confience . . ." and . . . "others with simple, credulous faces. . . ." apply so aptly to patients we see in any hospital casualty department, but never record in that manner.)

- G. K. Chesterton: *The Wisdom of Father Brown* (to realise how dated his prose has become).

- Mark Twain: *The Adventures of Tom Sawyer* (for descriptive elements).
- G. Guareschi: *The Little World of Don Camillo* (for the creation of emotive situations in few words, and humour with economy). Translated by U. V. Troubridge. These are a series of short stories in one book.
- Paul Gallico: *The Snow Goose and The Small Miracle* (for depth of feeling for words).
- James Thurber's short stories (for sheer fun).

The list is short and a personal selection, an "entree" to a new world. Nearly all are available in cheap editions which can be bought at a railway station before starting a journey of exploration of words. John Canning's "100 Great Books" (1966, Odhams, London) were picked for literary merit, original thinking and spiritual insight. Five to ten pages of commentary by various writers, are provided for each book so that the reader may flavour its content and enjoy the original the more. These 100 will be enough for a decade of reading. They are probably all available from your public library, so won't cost you a penny. Why not make use of this facility?

REFERENCES

Alchemist Glossary of Medical Terms (1963) 2nd edition *Alchemist*, London.

Alexander, R. S. (1953) Trends in authorship. *Circulation Research, 1*, 281–283.

Allen, E. M. (1960) Why are research grants disapproved? *Science, 132*, 1532–1534.

Armitage, P. (1971) *Statistical methods in medical research*. Blackwell, Oxford.

Ashe, G. (1972) *The art of writing made simple*. Allen, London.

Asher, R. (1972) Edited by Sir Francis Avery Jones. *Richard Asher talking sense*. Pitman, London. (Have a personal copy to mark the passages that amuse you).

Bailey, H. and Bishop, W. J. (1944) *Notable names in medicine and surgery*. Lewis, London. (Short essays on medical history).

Barber, A. S., Barraclough, E. D. and Gray, W. (1971) Closing the gap between the medical researcher and the literature (on Medlars). *Brit. Med. J., 1*, 368–370.

Bernard, C. (1957) *An introduction to experimental medicine*. Transl. H. C. Greene. Dover, New York.

Beveridge, W. I. B. (1951) *The art of scientific investigation*, Mercury, London.

Booth, J. D. L. (1973) Editor. *Directory of grant-making trusts*. Third compilation. Charities Aid Fund of National Council of Social Services, Tonbridge.

Braithwaite, R. B. (1968) *Scientific explanation (a reprint of a series of lectures in 1946)*. Cambridge University Press, London.

Bronowski, J. (1960) *The common sense of science*. Penguin, Harmondsworth.

Bronowski, J. (1964) *Science and human values*. Penguin, Harmondsworth.

Calnan, J. and Barabas, A. (1972) *Speaking at medical meetings*. Heinemann, London.

Calnan, J. and Barabas, A. (1973) *Writing medical papers.* Heinemann, London. (details of writing results of research).
Castle, W. M. (1972) *Statistics in small doses.* Churchill, London, (Highly recommended).
Cannon, W. B. (1945) *The way of an investigator.* Norton. New York.
Castiglioni, A. (1947) *A history of medicine.* 2nd edition. Knopf, New York.
Civil Service Department (1972) *Design of forms in government departments,* 3rd edition. H.M.S.O. London.
Clarke, E. (1971) *Modern methods in the history of medicine.* Athlone Press, London.
Cochran, W. G. and Cox, G. M. (1957) *Experimental Designs,* 2nd edition. Chapman and Hall, London.
Cooper, B. M. (1964) *Writing technical reports.* Penguin. Harmondsworth.
Cox, D. R. (1958) *Planning of experiments.* Wiley, New York.
Dorlands Illustrated Medical Dictionary (1965) 24th edition. Saunders, London.
Drucker, P. F. (1964) *Managing for results.* Pan, London.
Drucker, P. F. (1970) *The effective executive.* Pan, London.
Drucker, P. F. (1972) *Technology, management and society.* Pan, London.
Elder, J. D. (1954) Jargon – good and bad. *Science, 119,* 536–538.
Ellis, G. (1972) Edited by *Units, symbols and abbreviations: a guide for biological and medical authors.* Roy. Soc. Med., London.
Fisher, R. A. (1951) *The design of experiments* 6th edition. Oliver & Boyd, Edinburgh.
Flamant, R. (edited by) (1970) "Controlled clinical trials" International Union against Cancer: technology report series, Volume 7.
Fletcher, C. M. (1973) *Communication in medicine.* Nuffield Provincial Hospitals Trust, London. (Contains a fairly complete list of references on the subject).
Garrison, F. H. and Morton, L. T. (1970) *Medical Bibliography* 3rd edition. Deutsch, London.
Gelfand, M. (1968) *Philosophy and ethics of medicine.* Livingstone, London.

Hadfield, S. J. (1958) *Law and ethics for doctors.* Eyre and Spottiswood, London.
Hill, A. B. (1962) *Statistical methods in clinical and preventive medicine.* Livingstone, Edinburgh. (Must be read if only to realise how clinical trials and field trials work).
Hill, A. B. (1971) *Principles of medical statistics* 9th edition. Lancet, London. (Easy to read).
Himsworth, H. (1970) *The development and organisation of scientific knowledge.* Heinemann, London. (A well written biography of a thinker and doer).
Huff, D. (1958) *How to lie with statistics.* Gollancz, London.
Jones, V. (1974) *Creative writing.* Teach yourself books, London.
King, L. S. (1971) *A history of medicine.* Penguin, Harmondsworth.
Leach, G. (1970) *The Biocrats.* Cape, London.
Lockwood, A. (1967) *Diagrams: a visual survey of maps, charts and diagrams for the graphic designer.* Studio Vista, London. (Do look it over if only to realise the international nature of diagrams).
McGregor, D. (1960) *The human side of enterprise.* McGraw-Hill, New York.
McLachlan, G. (1967) From medical science to medical care. (Public Policy, the Resultant of Private Policies). *Lancet, 1*, 629–631.
MacNalty, A. S. (1961) edited by, *The British Medical Dictionary.* Caxton, London.
MacNalty, A. S. (1965) edited by, Butterworths *Medical Dictionary.* Butterworths, London.
Maxwell, C. (1973) *Clinical research for all.* Cambridge Medical Publications, Northampton. (Mainly about clinical trials).
Merritt, D. H. (1963) Grantmanship: an exercise in lucid presentation. *Clinical Research, 11*, 375–377.
Mitchell, J. (1964) *How to write reports.* Fontana, Glasgow.
Moroney, M. J. (1965) *Facts from figures.* Penguin, London. (Recommended reading as an introduction to statistics).
Morton, L. T. (1971) *How to use a medical library* 5th edition. Heinemann, London. (Essential reading if you are to get the most from a library).
Morton, L. T. (1974) *Use of medical literature.* Butterworths, London. (Excellent source book for journals in most medical specialities).

Oliver, R. (1971) *Principles of the use of radioisotope tracers in clinical and research investigations*. Pergamon, Oxford.

Pappworth, H. M. (1967) *Human guineapigs*. Routledge and Kegan Paul, London.

Parkinson, C. N. (1958) *Parkinson's Law, or the pursuit of progress*. Murray, London.

Potter, S. (1947) *The theory and practice of Gamesmanship or, the art of winning games without actually cheating*. Hart-Davis, London. (Out of print but worth reading. Some ploys are dated but still apply).

Reichmann, W. J. (1961) *Use and abuse of statistics*. Methuen, London.

Research Funds Guide (1968) British Medical Association, London.

Roberts, F. (1971) *Medical terms, their origin and construction*. 5th edition. Heinemann, London.

Scholarship Guide for Commonwealth Postgraduate Students (1975–77). Association of Commonwealth Universities, 36 Gordon Square, London (For those wishing to work outside their own country).

Scientific Research in British Universities and Colleges (1973–74). Volume ii Biological Sciences. H.M.S.O., London. (Guide to subjects of research in various institutions).

Sigerist, H. (1971) *The great doctors*. Dover, New York.

Speller, S. R. (1973) *Law of doctor and patient*. Lewis, London.

Stacey, N. A. and Wilson, A. (1963) *Industrial marketing research: Management and technique*. Hutchinson, London.

Stedman's Medical Dictionary (1972) 22nd edition. Williams and Wilkins, Baltimore.

Stein, E. B. (1971) edited by, *Dictionary of abbreviations in medicine and related sciences* 3rd edition. Balliére, Tindal and Cassell, London.

Stone, K. (1966) *Evidence in Science*. Wright, Bristol. (Essential reading).

Strauss, M. B. (1968) *Familiar Medical Quotations*. Little Brown, Boston. (Useful for the opening remark of a lecture on research).

Thornton, J. L. (1971) Medical libraries and research. *Brit. J. Hosp. Med.*, October, 485–490.

Townsend, R. (1970) *Up the organisation (or how to stop the corporation from stifling people and strangling profits).* Joseph, London.

Turner, R. (1973–75) editor. *The Grants Register.* St. James' Press, London and St. Martins Press, New York. (Lists scholarships in many subject. Published every two years).

Welbourn, R. B. (1966) The training and education of a surgeon. *Proc. Roy. Soc. Med., 59,* 934–936.

Wilson, E. B. (1952) *An introduction to scientific research.* McGraw-Hill, London.

Witts, L. J. (1964) edited by, *Medical surveys and clinical trials.* 2nd edition. Oxford University Press, London.

Wostenholme, G. E. W. and O'Connor, M. (1966) editors, *Symposium: ethics in medical progress.* Churchill, London.

REJECTION BY THE EDITOR

One well-known and popular journal rejects 75% of all material submitted for publication. Why? Mainly because the content was not scientific, but here are the seven commonest reasons:

- The investigator failed to show that the two groups in a clinical trial were comparable, on average; he failed to test a new drug against the best available drug, preferring to use a placebo instead, which is often unethical because the control group did not receive proper treatment.
- The failure to standardise methods was particularly evident in multicentre trials. (Remembering that the most difficult thing in the world to do is to make an exact observation, this lapse can hardly be forgiven).
- From the writing alone it was evident that the statistician had been called in too late to treat the data and make some result out of the mess. Not only should the statistician be consulted early but, by making a significant contribution to the paper, he is entitled to be an author.
- Retrospective studies rarely yield scientifically valid results because the material is biased and usually the findings are inapplicable. A retrospective survey may suggest an idea for future investigation, but the written document should not see daylight.
- Commonly, there were faults in presentation rather than in language, for the authors had not looked at a copy of the journal of choice to study the lay-out required. Where a summary is required this should be written with care, to state what the paper is about, what was found and and what conclusions were arrived at.
- Rejection will inevitably occur if the offering is too long and written in diffuse language, is a rehash of what is already well-known, or there has recently been a similar paper in another journal.
- Some papers are rejected because they would be better in

another journal, perhaps one with the required specialist readership.

The editor's job is to satisfy his reading public first, while still encouraging prospective authors to submit their offerings. He also tries to balance between providing sound medical information and sheer entertainment. Originality, therefore, comes high on the list whether this is saying the same old thing in a new way, seeing it from a different angle, making it controversial, or simply reporting new methods and new syndromes. Topicality and striking illustrations, with scientific validity, will always win a place too.

When your gem is rejected, what should you do? There is a choice:

- Rewrite the whole thing and resubmit it to the same journal. If the anonymous referee who has read and commented on the paper suggests minor changes, then these should be carried through quickly. If there are major alterations required, you will have to ask yourself, "Is it worth it?"
- Amend the manuscript slightly and send it to another journal.
- Consider modifying the length and content of the whole paper, noting the criticisms of the referee and submitting it to another journal.
- Withhold the paper until you have more data, often the wisest choice, but the most difficult to make.
- Contest the editor's decision, usually fruitless and not generally recommended.

REPORTS

At the end of the day you have to account to the members of the grant-giving body, how you have spent their money. If you have been thrifty, say so: they will be pleased and look favourably on you. The report may be only for the director of your institution or others but you will have to make a report – of failure or success, perhaps only a moderate offering. How you set it out *does* matter.

Many doctors do not enjoy writing reports; they consider them time-wasting procedures when there is so much practical work to be done in treating patients and in the research laboratory. This is a mistaken view. A well-written factual report may become a valuable document for the author and the receiver. As Cooper (1964) pointed out, it is the crucial test of a person's knowledge to be explicit and articulate, to make it intelligible to others: "This is surely the difference between the technologist and the technician". The fact that a research worker looks upon report writing as the least important and least necessary of his activities, is no excuse for doing it badly.

The value of a report is related to the cube of the work force. If there are only two researchers, a report is unnecessary and ordinary conversation will do. If there are 60 people to inform then a report may be the only way of keeping in touch. Indeed, it can be circulated widely and studied at a convenient time; with its terms of reference precisely defined, thoroughly prepared and checked, a report may lessen the likelihood of premature and hurried judgements.

Composing a report on a meeting can be an exacting task. One usually makes notes at the time, so that all the information needed and actions recommended are to hand, but even so the scribe will have to have an analytical mind. It is also advisable to write the complete report as soon as possible afterwards. Surgeons, after all, make a practice of writing the details of an operation immediately it is finished (or dictate their findings into a machine); they don't wait until the end of the day because memory is too fickle and fleeting for that.

The writing of a report differs from the writing of a paper for publication in three important aspects – the specificity of its content, the known readership and the limited time for preparation. Reports are sometimes demanded and so those who ask have to read them. You, therefore, have captive readers. If writing a report is an unwelcome chore, reading it is often an unnecessary bore. Great literary style is not expected but brevity and clarity are appreciated. Even if the readership is known in advance, the interest will not be uniformly shared. Those who will be most affected by your report will read in detail and probably talk it over with you later; those least affected will scan it quickly and, if it is difficult to understand, will file it away without being any the wiser. Before putting pen to paper, ask yourself: Who is it for? What is it about? How long should it be? Is it a record of work done, as the basis for further experiments? Is it solely to provide information, to circulate new ideas, to cross-fertilise ideas, to indicate future action?

In writing a report there are three stages:

- Ideas are jotted down quickly on a piece of scrap paper. It is important to think of what to report, what relevant facts need to be known, what headings.
- The rough draft follows, usually in a logical order.
- The final polish and the addition of interest, with the reader in mind. The third stage should not be tackled immediately but after an interval (of 10 minutes, but better a day or two if time permits). Now is the time to rephrase sentences, to eliminate jargon and to question the exact meaning of the words written. Accuracy in statements, and particularly in numbers and statistics, is terribly important. Assemble all the facts, not just your own, and assume that your reader is busy and won't be able to study your report in a quiet room, at leisure. Reports can be good, bad or indifferent. But on the written word you will be judged fit to receive or be denied further funds. So take good care how you write and what you write. Kindly editors spend much time correcting scripts and turning them into something readable for publication; grant-giving bodies do not.

A report is a written statement of a situation and so must include:

- the facts collected,
- how they were obtained,
- their significance, and,
- certain conclusions or recommendations.

There are two general principles which apply to the writing of all reports. These concern the structure and the layout.

STRUCTURE OF A REPORT

A report conveys specific information to a specific reader and requires a structure like any other piece of writing. Because time for its preparation may be limited, the first draft may have to be the final composition. It is therefore doubly important to write concisely and clearly and to use simple, familiar words in uncomplicated sentences. The text may be simplified by keeping to the essentials only, in language which the reader will understand. Statements will have to be justified by facts and figures, so quantify and avoid generalisations (so state dimensions, amounts, weights and volumes – and do avoid the use of percentages for small numbers). If you think that your report will require about 6,000 words, try to reduce it to 3,000 and put the supporting information that is really necessary in an "appendix".

The plan of the report might be as follows:

- The title page, with your name, the date of the report and for whom it is intended (its circulation).
- If the report is more than 5,000 words, a contents page is a good idea.
- The next page should be an abstract if the report is longer than five typed pages. This should contain all the essential facts and be so constructed that it can be circulated alone to other interested parties as the article in miniature.
- The introduction should tell the reader why the investigation was made, how it was carried out and what were the chief findings. You do not have to capture the reader's attention as for a published paper, but you have to hold it. The introduction should, therefore, be simple, short and direct.

- The main body of the report requires an orderly presentation. A chronological sequence is ideal for a medical report on a patient but unsuitable for a research report, even though the work was done in a chronological order. It is important to decide on the most appropriate format and to arrange the observations within it before constructing the prose. One should deal with simple facts and ideas before recording the more complex, so that the reader is led logically on.
- Interpretation may not be called for if the facts speak for themselves. If they do not, then an hypothesis to explain them is required; since this is often a personal interpretation, the use of "I" or "we" is in order. There is a curious notion that objectivity can only be obtained by writing impersonally. I entirely disagree and would argue strongly for personal involvement of the writer in any report. I also disagree with the notion that vagueness is in some way the same as being diplomatic.
- The main recommendations constitute the most important part of the report for the busy reader. If the main facts and arguments have been set out well, he has a simple decision to make; either to agree or disagree. If there are no recommendations then this section should contain the conclusions, cogently expressed.
- References, if any.
- Appendices, if required. These should contain extra detail for the selective and interested reader, which do not form an essential part of the main report.

THE LAY-OUT

Layout is important: it should not be left to the secretary who types or stencils the report. An attractive appearance to a report tells the reader immediately that you have taken trouble. Furthermore, a durable cover allows the work to be handled frequently and passed to others without looking unkempt. The cover should be transparent to allow the title to be seen or, if opaque, of a distinctive colour with the title boldly set out. A faceless report quickly becomes lost on anyone's desk. The following are important for good appearance and ease of reading.

- Each new section should start on a new page.
- Sub-headings should be used frequently so that the reader

can consume a small amount and perhaps a new idea, piecemeal. He will also follow the structure of the report more easily.

- Paragraphs should be used discretely to break up the typed page. A sheet of solid type is hardly conducive to detailed reading.
- Tables should be inserted so that they can be seen at the right time within the script. If there are about 250 words to a page of type, it is not difficult to ensure that tables are inserted in the correct place.
- Diagrams and graphs, drawn in black ink, should be used and be continuous with the script if they facilitate the reader's understanding of your message and allow the script to be shorter. When the report is to be duplicated it is usually easier to have the illustrations photocopied rather than draw them on a stencil sheet. Alternatively, they can be photocopied, trimmed and mounted on the typed sheet.
- If you produce artistic and neat work, your appreciative reader is more likely to look it over again and so better understand what you have written.
- It is advisable to leave good margins at each side; on the left for binding, on the right to allow for notes made by the reader (if he can make notes at the margin he has become involved and that's what matters).

Remember that reports to grant-giving bodies are often published. Hence your reputation may be assessed by others on the quality of your written contribution alone; so write better than others, using well-chosen descriptive words and make the report interesting and exciting because your work will be in competition with that of other grant-holders. People want to know what happened, who did it, what was found, what is the next step in any research. It is sometimes justifiable to give an opinion on the results and the progress of a project, but too often this is not borne out by subsequent events.

Finally, all reports should conform to the 5 C's rule – clear, consistent, complete, concrete and correct. Good books to consult are those by Cooper (1964) and Mitchell (1974): see REFERENCES. Try not to produce the kind of report, condemned by Garland (1952. New England J. Med. 246. 801).

REQUIREMENTS OF RESEARCHER

Does devotion matter? Of course it does, for without it there is no spiritual incentive. The urge to help another human being in suffering cannot be suppressed by purely financial gain.

If you pursue a craft then you must be an expert: no amount of research will make up for that. Indeed, if you are a research surgeon then the quality of your surgery in clinical practice must be above average, and preferably better than anyone else's. Only then does your research mean something for clinical practice. Research is much harder than clinical work, more exact with less opportunity to cover up mistakes. It is not by coincidence that those centres of medical excellence are also the very centres from which the best research emanates. Excellence breeds success, but in the end excellence depends on the personality, the drive, the integrity of the individual. "If you can't do it excellently don't do it at all" (Townsend, 1970). Because if it's not excellent it won't be profitable or fun, and if you're not in research for fun or profit, what the hell are you doing here?

There are probably 9 qualities which are required:

- Average intelligence: the highly intelligent so often are unable to pursue one objective to a successful conclusion and tend to explore interesting, but non-productive bye-ways.
- Simple curiosity is at the root of most research and success so often depends on asking "why?".
- Intense pre-occupation is obviously desirable when there are so many distractions in daily life and fortunately it can be learnt, but not without effort.
- Showmanship: that's pretty obvious, and is equivalent to self-confidence.
- Some selfishness, so that the work can get done.
- Most successful research people have a certain streak of ruthlessness, both with themselves and with others.

- The ambition to do well, to finish the job, to do better next time.
- The ability to recognise that compromise is essential, sometimes.
- Motivation, that is, the intense desire to do research, although placed last is probably the most important attribute. Many people say they want to do research but lack the drive to do it; those who have, will.

Simple observation indicates that most people are slightly uncertain of their ability, capacity for responsibility and judgement, but react well to encouragement; wars, disasters and crises all demonstrate this to be so. The hidden courage of ordinary folk always surprises, yet these are the very people who make ventures successful for others. In detailed analysis they will be found to have more than one or two of the qualities listed here. Lastly, do please note that "nice guys finish last", so a certain amount of steel is valuable. Indeed, the inner compulsion to do research and a certain attitude to society and people are as important as anything else, but in the end, if there is no professional efficiency the other qualities do not count for much.

RESEARCH DEFINED

The time has come to define what is meant by research. It is an investigation or enquiry, usually made closely and carefully, to discover some fact or truth about the nature of the universe and all who inhabit it. Human research is limited to the study of Man in health and disease, for advancement of his well-being. Because much research is comparative the study of healthy beings is as essential as the study of the sick, whether the measurements are made in a test-tube, on animals or in Man. Research, to be effective, must be sincere. It's the first requirement. Sincerity means being ourselves, saying what we really mean.

The way of the investigator depends on using the scientific approach combined with certain well-tried methods of measurement, which do not exclude the development of analytical methods as part of research. The philosophy of research is therefore fairly clear cut, but if the scientific method cannot be applied (and there are several areas where it cannot yet) then investigation either is not possible or becomes pointless. The gathering of observations or facts alone, without an hypothesis in which to fit them, does not constitute research.

It used to be thought that there were two forms of research – fundamental and applied – but recently the differences between them (if they ever really existed) have become blurred. All new knowledge will be put to use sooner or later. What is becoming more and more evident is that research in one subject, where its application in one area is not immediately evident, often will be accepted for use in another. In technology this is known as "spin-off"; an incentive to every researcher to look to other fields for possible solutions to his own problems, to save time, money, effort and thought.

Advances in medical management are almost entirely due to new inventions rather than to modifications of the old. The pattern of invention goes like this:

concept
↓
trial and error
↓
technical achievement
↓
acceptance by majority
↓
action in practice

The first three stages might be designated "fundamental research" and the last two as "applied research" but the separation is rather artificial. Many people collect records of patients for clinical research. They are likely to be disappointing unless:

- records are kept with an end in view,
- they are part of a planned experiment,
- the observations recorded are designed to withstand critical analysis and provide a specific answer to clinical problems.

Without these three objectives in mind, a cabinet full of interesting cases will form the basis for a book of reminiscences but never a published paper on research.

REWARDS

The reward of research is not fame or success, but intoxication; that is why so many researchers, good or bad, are unable to give it up. Lectures, publications and recognition are almost by-products even though they may have been the motivating force to do research in the first place. A curious situation indeed.

SALARIES

There is no such thing as equality in research, not even equal opportunity because opportunity has to be devised. Many people in a research team may receive the same money for work done (or not done sometimes) yet a way has to be found to reward competence, ability and productivity (by productiveness I mean results and not the effort put into the job; they are often related but not always). The rewards can be in the form of status, time off to do other things, or extra cash.

It is therefore important, when employing them for research, to start people at the highest level comparable to their apparent abilities, but have freedom to promote within that salary scale. Everything in life has to be paid for, even experience can be costly and, on the whole, we get what we pay for. If penurious, then do not expect top-notch assistance; so reduce the complexity of the research project until such time that it shows promise and some results, when it may be possible to apply for further funds to obtain better service.

In employing assistants remember three things:

- Try for the best available person for the money.
- Fight for increases in salary for worthy staff and never put pen to paper for those unworthy.
- McCormick's Law applies (comparable to Parkinson's Law – see People) which states that "however large the salary to start with, the total amount shrinks to accomodate the bare necessities of life" – and this has nothing to do with taxation! (McCormick J. 1975. Motives in Medicine. *Proc. Roy. Soc. Med.* 68. 16–20).

SAMPLE SIZE

Most statisticians run for cover as soon as they are asked about the size of samples. Yet the researcher wants to know: "How many subjects do I need to investigate to obtain a statistically significant result?". In setting up a clinical trial the answer to this question is very important and yet it is not usually easy to provide. The difficulty lies in deciding what degree of precision to aim for in the estimates of the variables and their differences. Contrary to expectation, the onus of this decision rests with the researcher and not the statistician who can only provide the methodology, but not the data necessary for the solution.

Let us consider the problem of comparing the means of 2 populations, assuming that they have the same standard deviation, σ, and that 2 equal random samples of size n are to be taken.

The following method is often used to obtain an estimate for n. The investigator first decides on a value δ which represents the size of difference between the treatments that he regards as important. The value of δ will vary with the nature of the experiment. He also decides on the significance level of the test, that is, the probability associated with the error of concluding that a difference exists between the means when in fact, it does not. Significance levels are arbitrary, but popular values are 5, 1 and 0.1 per cent, such that a 5% significance level indicates that 1 time in 20 the difference observed will occur by chance alone. Then, using the fact that the standard error of a mean is given by

$$\frac{\sigma}{\sqrt{n}}$$

it can be shown that to satisfy the stated criteria we must have

$$n > 2\left[\frac{U\sigma}{\delta}\right]^2$$

where U is obtained from tables and is the standardised normal deviate for the appropriate significance level. For the 5% level U = 1.96: for the 1% level U = 2.56. Substitution in the right hand side of the equation yields a value for n. The difficulty arises in the specification of δ and the assumed knowledge of σ, the population standard deviation. In fact σ is usually unknown and has to be estimated. A pilot or preliminary sample may be drawn from the population and the standard deviation, computed from this sample, may be used as an estimate of σ. Observations used in the pilot sample may be included as part of the final sample. Alternatively, estimates of σ may be available from previous or similar studies.

The same principles will apply in the related problem of comparing two proportions or percentages. For example, the proportions may be the incidences of deep vein thrombosis after coronary heart disease in 2 populations. The δ is then the difference in proportions that the investigator regards as important and σ the combined standard deviation. If p_1 is the percentage of population 1 with D.V.T. then q_1 is the complement of p_1 (that is the percentage in population 1 without D.V.T. because $p_1 + q_1 = 100$). Similarly if p_2 and q_2 represent these percentages in population 2, then σ may be taken approximately to be the pooled value

$$\sigma = \sqrt{\tfrac{1}{2}(p_1q_1 + p_2q_2)}$$

and substitution in $2\left(\frac{U\sigma}{\delta}\right)^2$ indicates the minimum value for n to yield a significant difference in the proportions at the appropriate significance level.

So once again the investigator has to specify his significance level, provide a value for δ and he will probably have to estimate σ, the standard deviation. This in turn implies estimation of p_1 and p_2 which may be the consequence of a pilot survey or available from published work.

For example, suppose $p_1 = 30$, $q_1 = 70$

$$p_2 = 40,\ q_2 = 60$$

$$\text{then } \sigma = \sqrt{\tfrac{1}{2}\,((30)(70) + (40)(60)}$$

$$= 36.12$$

Suppose the investigator regards a difference of 15% in the incidence as important and he is interested in detecting a difference at the 5% significance level.

$$\text{Then } n > 2\left(\frac{1.96 \times 36.12}{15}\right)^2$$

$$n > 44.5$$

Thus the investigator should take 45 subjects from each population to form his sample.

This approach to estimating the sample size in an experiment, whether applied to quantitative data in the form of means or qualitative data in the form of percentages, is only one of a number of approaches. The final inequality determining n will vary according to the specification of the problem. However, the principles rely on standard statistical theory and thus, given expert knowledge in that field, they can always be applied. It should be emphasised that this is an estimation procedure, not an exact science, and the final solution is dependent on prior knowledge and specifications which are the responsibility of the investigator. Estimating a sample size is not easy, if only because, when the "guesstimate" is wrong, so will be everything else.

SCIENTIFIC METHOD

As Sir James Jeans once wrote:

"Science fishes in the sea of reality with a particular kind of net called the scientific method, and there may be much in the unfathomable sea which the meshes of science cannot catch". Lord Kelvin also recorded: "When you can measure something you begin to understand it". Exact measurements in medicine are often difficult and do not allow for the natural biological variations between individuals. One method of measurement is by comparison – and the controlled clinical trial is a good example. Even so, we have to understand what is meant by Scientific Method.

It is sometimes forgotten that the scientific method as we know it today is still quite new (see any recent book on "management"), the product of the scientific revolution which lasted for well over three hundred years. The application of the scientific method to problems in medical and related subjects has a freshness that is nowhere near mature, for it provides an attitude which is realistic as opposed to ritualistic.

Science in medicine is an expanding framework of exploration; it is neither the cultivation of special techniques, nor an acute appreciation of the wonder and unexpectedness of the diseases we see and treat.

THE SCIENTIFIC REVOLUTION

Following the almost complete ablation of knowledge in the Middle Ages, the Renaissance of the 16th century saw the rebirth of literature and pictorial art. The scientific revolution began a little later as part of the Italian Renaissance, when natural history and scientific biology were stimulated to rediscover Greek sources of science and to re-examine them.

It really began with Andreas Vesalius (1514–64) who substituted the trained eye of a great anatomist for the historical hearsay of Galen (129–199), by material in the dissecting room and recorded in *De Humanis Corporis Fabrica* (1543). Harvey's discovery of the circulation of the blood in 1628 followed quite naturally, and aroused the first serious conflict between ancient and modern theories of physiology. Harvey did for physiology what Vesalius had done for anatomy. Until this time the only instrument for research was formal logic: men were educated as logicians and not as observers.

The scientific revolution was a reaction to medieval scholasticism, yet scholasticism contributed to it. It was a reaction to the traditional science embalmed in books, which has continued to this day (see textbooks of surgery, if no other), so that intellectual subtlety which has minimal relation to things as they are is no longer held in such high esteem.

The revolution did not end until the classifications of Linnaeus. In the 400 years since the scientific revolution we have learnt that we gain our ends only within the laws of nature. "Man masters Nature not by force but by understanding" (Bronowski, J (1964) Science and human values).

Although medicine has emerged from the natural history stage its methods have not, and perhaps can never be, identified with those of physics. For, however great its development within laboratory walls, there must always be room for those techniques of description and discrimination with which the naturalists of the 16th century first strove to create a science.

THE SCIENTIFIC METHOD

The four stages of enquiry which constitute the scientific method are:

- Observation
- Hypothesis
- Prediction
- Verification

In experimental medicine this is frequently modified to:

- Observed facts
- Hypothesis
- Controlled experiment
- Interpretation of experimental results, which then become another set of observed facts.

The essential thing is that we move in an orderly manner from 1 to 4, and then start all over again. The sequence of observation – hypothesis – controlled experiment – interpretation is followed exactly and in this manner results which mean something are obtained. Science, then is tested knowledge. The scientific method is merely the means for testing, for it is explicit and systematic.

1. *Observation*:
We start with some observation of facts – not just a single fact but more than one. (It must be understood that the observation "it is dark" has little meaning or value: but "it is dark at midday" allows several possible hypotheses.)

Science deals with things that can be observed and not with impressions that can be felt in a spiritual sense. For this reason science is impersonal, so that personal speculation in situations where there is little knowledge is avoided. Hence psychiatry, on the whole, is a closed subject – in spite of many worthy publications.

Knowledge grows as more information (observed facts) becomes available, and so hypotheses change. This unstable, restless state is understandable. As we learn more we can correct our concepts so that they fit our observations better. It is doubtful if any scientific discovery has been made by accident. Certainly an observation may come as a result of an accidental happening, but the observation itself is no accident: it is the product of an alert mind. "Chance favours the prepared mind" said Louis Pasteur when asked how he had managed to make so many discoveries.

The discovery of penicillin was no accident. The fact that penicillin spores contaminated some culture plates was an accident, which has happened before and since; it was the positive observation of the zone of no growth surrounding the spores that set Sir Alexander Fleming apart from others as an observer. This points out too, the fundamental difference between "seeing" and "observing".

Observation, then, appears simple and straight-forward until one realises how many errors of observation have been made in the past. An observer observes what happens, collects and studies facts relative to the problem he is engaged in.

2. *Hypothesis*

Science and technology are not just a collection of facts. Facts call for judgement and interpretation – this is the function of an hypothesis. The hypothesis must have one vital characteristic – it must be testable. It does not matter how extraordinary the hypothesis may be, if it can be tested by controlled experiment it is legitimate. Most hypotheses come straight out of the minds of men; they are rarely contrived.

The first and most obvious test for a hypothesis is a mental one – is it consistent with the observed facts? The history of science has shown that when adherents of a general concept have to retreat to a position of pleading special cases, defeat is imminent.

Often a new hypothesis can only become established when general advances in knowledge and technical ability make it possible to test it accurately. Hence, the germ theory of disease (which started with Hippocrates?) had to wait for the microscope. Advances in science depend on previous advance in knowledge and technique: they are not isolated events. The irrisistable need to explore is the scientific spirit.

The scientific attitude encourages people to put forward their own points of view: but it insists that they should support them by reasons which other people can verify, and that they should be willing to accept the judgement of critical experiments whether they have made out their case. In other words, it is a recognition that one belongs to a community in which one should do one's damnedest to pick holes in its beliefs. There is no better association between freedom and order which also retains the full values of both.

The hypothesis, then, is an explanation for the facts that have been observed and is a natural consequence of observation and investigation. Following most observations there is usually more than one hypothesis; the number is only limited by the imagination of the observer. Since it is really only possible to proceed to the next stage with one hypothesis at a time, it is common to select the hypothesis that first entered one's head or seems the most logical. Facts call for judgement and interpretation – an hypothesis.

3. *Prediction*
From an hypothesis it now, for the first time, becomes possible to make deductions. If the hypothesis is satisfactory, new knowledge, not known empirically before is born. When an hypothesis receives continued support it becomes hallowed and is then known as a "theory".

4. *Verification*
Further new facts are collected to test the predictions made from our hypothesis. With this step the cycle starts all over again.

The object of the scientific method is to uncover the secrets of nature in an orderly manner. Unfortunately, there is no final truth in science, for although failure to refute an hypothesis may increase confidence in it, no amount of testing can literally *prove* that it will always hold. We seek "the truth, the whole truth, and nothing but the truth" but as Osler says, we may have to be content with "broken portions" of the truth. The scientific method is a discipline which helps to fit "broken portions" together even though we may not appreciate the whole.

LIMITATIONS OF THE SCIENTIFIC METHOD

● Science is a way of describing reality. The scientific method therefore is confined by the limits of observation, since it asserts nothing that is outside observation. Anything else is not a science, it is scholastics. Because observation is the prime stimulus to the sequence in scientific method, the correctness and accuracy of the observed facts is of paramount importance. When the initial observation is inaccurate it is no wonder that the subsequent hypothesis turns out badly.

Agnes Arber (1954) recorded the biologist's road to reality and described six stages in the journey along the way:

1. choosing a question
2. collecting the data
3. interpretation
4. testing the validity of any conclusion
5. writing up the work done
6. contemplation of the work in its wider context.

● Science asks questions of nature but we have no method to

ensure that the right question be asked, except by trial and error.

In many scientific problems the difficulty is to state the question rightly – once that is done, it may almost answer itself. One of the most notable characteristics of the scientific approach to a practical problem is an insistance on deciding exactly what one is trying to do; it is surprising how hard it often is to decide this.

- Normally only one question can be posed at a time, but nature frequently confuses us by demonstrating that several questions have to be answered concurrently. Modern statistical techniques have been developed to contend with this situation. The method of testing knowledge step-by-step is tedious and hence sudden changes in concepts do not occur.
- There can be no proof. What we have learned from the past is knowledge only because the future shows it to be true. Science accepts what works and rejects what does not.

The major objection to the scientific method is to its fundamental uncertainty. The desire for an ultimate bedrock of stability is very important in man's nature and science can never supply such stability. The standard scientific methods for dealing with things were worked out for simpler situations than some of those found in medicine; and when applied to objects which are both complicated and highly organised and unified, they tend to yield a lot of isolated facts about details (but to let slip the secret of the thing as a whole, which is what one is after). It is an orthodox and respectable scientific technique to measure things, if necessary to measure two or more things, and find out how well the measurements fit. But it is also essential to have some measurement of the "goodness of fit".

ADVANTAGES OF THE SCIENTIFIC METHOD:
- It introduces two creative ideas – the idea of order and the idea of causes. A third, the idea of chance is dealt with under statistics.
- It resolves to some extent the ambivalence within all of us between fact and logic.
- The scientific method does not consist of the collection of facts alone; nor is it enough to think: "it is the union of empirical

fact and rational thought that gives it substance." It is a continuous to-and-fro movement of actual discovery, then back to thought about the implications of what has been found – a step-by-step progress of practice and theory.

- It high-lights that the essence of cause and effect is their sequence: the cause must come before and the effect must follow after, as John Stuart Mill has pointed out.
- It is an orderly language for describing some events and predicting others like them.
- It recognises that every prediction carries with it its own measurable uncertainty and has developed methods to assess this. Action, thus, is based on probabilities rather than on certainties.
- In this country, and many others, people value their liberty to do what they like regardless of what others may think. The scientific method supports this attitude, because in science the freedom to differ from the majority is one of the essential requirements of the work. Once this freedom, for any scientist to disagree with the majority including his elders and betters, is destroyed scientific progress would come to an immediate end. The scientific attitude allows one to believe in some things more, and in others less, but in none with such absoluteness that evidence becomes irrelevant.

Some comments on science have denied categorically that there is any such thing as a specific "scientific method" by saying, for example, that science is organised common sense. This is to deny the events of history. Scientific method, shaping itself to the needs of practising scientists, vindicated rather by results than by preconceived logical rigour, has remained an enigma to philosophers; its implementation in practice was more important than its implicit formulation.

Much more has been learnt about the nature of man in the last century and a half than in the previous 5,000 years. Of this there can be no doubt. From the nineteenth century onwards a scientific paper on almost any topic is intelligible as a direct precursor to research which still continues. For example, Alibert recognised keloid scars for the first time as a clinical entity in 1815. In 1961, Cosman and others wrote on the same subject and, indeed, had little more to contribute. But the two papers can be read side-by-side and make a perfectly comprehensive story of clinical investigation. We tackle today the same problems, but with stronger tools, as our forefathers did; the same insistence upon accuracy of observation, the same confidence in quantitative experimentation, and the same enmeshing of theory, hypothesis and factual reporting which medical philosophers then and now found so resistent to logical analysis.

A scientific approach to medical problems must be the sum of its many aspects – experimentation, mathematical analysis, quantitative accuracy, and so on – varying according to the nature of the problem. In this approach, mathematical reasoning is important because it is simple and avoids doubt and confusion; it is capable of infinite ramifications without falling into inconsistency; it is totally indifferent to the observer and his problems of perception since it cannot extend beyond what the seventeenth century called the "primary qualities". Hence the need for statistics in all biological phenomena.

SECRETARY

If you must have a secretary then pick a good-looking one and make sure she is worked hard. When there is a slack period, let her have the day off for shopping; the idle secretary only attracts people to come and chat when they should rightly be employed elsewhere. Anyway, the busy secretary always finds time for the important things to get done and soon becomes discerning, even if she did not start that way. Hence, she becomes a valuable asset to the research team by preventing the machinery becoming clogged with paper.

Townsend's approach, for the busy man, has much to recommend it. When engaged on laboratory work, the secretary takes *all* telephone calls and promises that you will 'phone back at a stated time. It is thus possible to deal with all callers during a one hour period, instead of having multiple interruptions throughout the day. To make the system work well, the secretary has to be tactful but firm, knowledgeable yet helpful. "A good secretary can think for you, act for you, anticipate every whim and increase your output – or ruin you" (Townsend again). A bad secretary is no more than a status symbol and should be sacked. Better still, explain the research to her and she will type the papers for publication more accurately, more intelligently and may even come up with a good suggestion.

A secretary is not essential for research – an aide at best, but a very convenient aide. The wise one will make her the public relations officer for the team, whence she may well excel because she has become a member of the organisation.

SELECTION COMMITTEE

A special form of committee is that set up to select a candidate for a particular post. Many people find this commission rather daunting; they do not know what to ask the candidate and often have a poor idea of the kind of person the post requires. There are 7 things to look for:

1. Education
2. Training.
3. Experience

(1–3: All should be available from the application form.)

4. Special qualifications for the post
5. Skills – obtained from references
6. Physical traits – that is medical fitness, which can be elicited at interview and confirmed by a subsequent medical examination which is compulsory for most employing authorities.
7. Personality of the applicant – which should become evident at the interview.

For members of the interviewing committee here are some suggestions. Every candidate should be put at ease and the normal courtesies shown. Usually, the chairman of the selection committee will tell the candidate the essential nature of the job, have a few general questions, listen carefully to the replies and, at the end of the interview, encourage him to ask questions about the work involved. It is the chairman's prerogative to leave a favourable impression on the applicant, as well as to sum up the decision of the committee at the end of the meeting.

- Avoid questions which merely ask for information which is on the application form. "What is your age?"
- Avoid questions which could measure qualities more easily by other means. "How good are you at . . . ?"
- Avoid needless, superfluous and irrelevant questions. "What is your religion?"
- Avoid suggestive questions which demand certain answers.

- Avoid questions based on wide assumptions.
- Never argue.

If this guide is followed and you have defined the objective before framing questions, you will quickly be able to make a preliminary assessment of each candidate. The final assessment is made when all candidates have been interviewed, when you have heard your colleagues' opinions, listened to glowing testimonials and discussed among yourselves. It all sounds very simple but, of course, it really is not. Selecting the right candidate is an art, not a science and mistakes become obvious only later.

S. I. UNITS

Having had to learn almost another language in order to take part in and to follow the research of others, we have now to accept a new code of measurements. These, an extension of the metric system known as the "Systeme International d'Unites (or S.I. units), have now become the internationally accepted units of measurement throughout industry and science. While metre, kilogram, seconds and amperes remain as before, the amount of a substance is a "mole" (mol.), thermodynamic temperature a "kelvin" (K) and luminous intensity "candela" (Cd). Multiples and subdivisions, such as 10^6 (a mega or M) or 10^{-6} (micro or μ), also have special symbols and large numbers are to be so spaced out so that commas will not be used except as decimal points – 4 000 000 or $4{,}0 \times 10^6$.

New units have also been introduced, such as the newton (N) for force (one newton is the force which accelerates a mass of one kilogram, by one metre per second squared) and for pressure (including blood pressure), the pascal (Pa) which is the pressure exerted by one newton acting on an area of one metre squared, whereas one joule (J) is the work done by the force of one newton acting over one metre.

All these units will be unfamiliar at first, but will introduce a worldwide standard for measurements. For most of us, a haem-maglobin expressed as g/de (the decilitre being 100 ml) and fasting plasma glucose as nmo 1/1 will be strange. To overcome these difficulties the booklets by D. N. Barron (*S.I. Units*, 1974, published by B.M.A., Tavistock Square, London) and by G. Ellis (*Units, Symbols and Abbreviations* 1972, Royal Society of Medicine, Wimpole Street, London) contain helpful information and conversion scales. Clinical Science & Molecular Medicine, 1974, (Volume 47. pp i–xvii, "Guidance for Authors) and the Journal of Physiology, 1966 (Volume 182, pp 1–33) contain similar information.

SIGNIFICANCE

Kenneth Stone wrote that "the beliefs of most people are based, not on evidence, but on testimony". The object of statistics is to provide a method for assessing the validity of the evidence, by introducing a value for the probability in each situation. In other words, what are the odds that the results of an experiment are merely due to chance, a fluke if you prefer the term?

The word "significant" is used, but it sounds more important than what it means. There are three levels of significance commonly used, or levels of confidence, expressed as 5%, 1% and 0.1% to indicate that the result could have occurred by chance 1 in 20 times, 1 in 100 or 1 in 1,000. The 1 in 20 is usually accepted as "significant" (and the P, or probability value is also given, $P < 0.5$) but for most clinical work it is better to set your sights higher, say at the 1% level because you will feel more confident when deciding two further questions which arise from the research:

1. Is it worth it? That is the cost-effectiveness of introducing a new therapy.
2. Does it save anything? Its cost-efficiency. Please note that significant is not the same as important and never has been.

A test of significance (and it must be specified which one: 't' test, chi-square, variance, or others) is essentially a rule for deciding, from examination of the data, whether or not to reject the hypothesis. The rule, or guide, for that's all it is, is made to satisfy two conditions that are clearly desirable:

1. Hypotheses that are true shall be rejected only very occasionally, and
2. Hypotheses that are false shall be rejected as often as possible.

The level of statistical significance attained measures the uncertainty involved in taking, say, an apparent difference between two treatments as real. Hence,

● Not statistically significant at the 10% level	= data are consistent with no (or zero) contrast. The new treatment is no better than the old (and may be worse)
● Statistically significant at or near the 5% level (but not at the 1% level)	= data give good evidence that there is a true contrast (or difference) between treatments.
● Statistically significant at or near the 1% level	= data give strong evidence that the true contrast is not zero and that, at last, you're on to something.

But again it must be emphasised that statistical significance is not the same as technical importance.

SOURCES OF INFORMATION

In order to do any piece of research with the slightest hope of publication (the natural end product, for how else can we let others know?) it is essential to inhabit a good library from time to time. By doing so you may find that:

- your project has been done before and very well, or
- it has been done before, but badly,
- the project should be modified in the light of new discoveries which have a direct bearing on your own work,
- someone else is actively working in the same field and should either be consulted or invited to join in a combined project.

There are really two problems – how to obtain a lead into the literature on your subject and how to keep up-to-date with current information. The main source of information will be a well-stocked library because most of the useful books are very expensive. In Britain, the libraries of the Royal Society of Medicine, the British Medical Association, and that of the Royal Postgraduate Medical School too, all provide good source books as well as allowing the member to draw on the National Lending Library (at Boston Spa) and other specialist libraries for books not otherwise easily obtainable. Subscription libraries such as Lewis' in London and Ferrier's in Edinburgh are helpful for individual texts. To get started, the researcher should consult in this order:

- Colleagues in the same institution or working in the same field (technicians and your chief). Hence the importance of frequent discussion among all members of a team before, during and after every research project. It is surprising how often a vital piece of information is disclosed suddenly by the most lowly and unassuming of people; the search for knowledge by asking questions pays off. If you have to stand on other people's shoulders to get a better view, do so, but remember to recognise in print this facility. The important thing is to pick the brains of anyone and everyone who might be able to help.

● Textbooks are easy to look up and may provide essential references which can then be consulted. However, a textbook tends to describe things in general terms and as such may not help to define a specific problem.

● The Index Medicus, probably the most useful and effective guide to medical literature. It is published monthly and cumulative annual volumes are also issued; review articles are in the front, key words are in alphabetical order with a set order for languages (papers in English come before those in French or any other tongue). About a quarter of a million papers from 2,200 journals are listed annually by title, often under several headings.

Looking up the appropriate word for your subject may be difficult. For instance, the word "adrenalin" does not appear, but by consulting MESH (Medical Subjects Headings) you will be told to look for "epinephrine" the equivalent word used in the U.S.A. where Index Medicus is published.

● Other useful sources are:

Medical and other dictionaries for exact definitions of words
Index catalogue of the Surgeon-General's library (until 1955)
Current medical literature
Current Contents (weekly – now in two sections – Clinical Practice and Life Sciences), and *British Medicine* (monthly)
Royal Postgraduate Medical School Library Bulletin (about 1,000 titles of 70–80 journals) – weekly
National Library of Medicine Catalogue
Current Catalogue (N.L.M.)
Lewis' lending library
Medical Directory (for an author, address, position and other papers)
Medical Register (issued by G.M.C.) and mainly for addresses
British Encyclopaedia of Medical Practice
British Surgical Practice
Modern trends series of various subjects (anaesthesia, medicine, surgery, etc.)
Recent Advances series of various subjects
Annual Review of Pharmacology

Annual Review of Physiology
Yearbooks (of various specialties)
● Abstracting journals:
Abstracts of World Medicine
International Abstracts of Biological Sciences
Surgery, Gynaecology and Obstetrics
Anaesthesia
Survey of Anaesthesiology
Excerpta Medica (in 32 separate subjects)
New England Journal of Medicine (weekly)
Plastic and Reconstructive Surgery (monthly)
● Interlibrary loans from:
British Library Lending Division (Boston Spa, Yorks)
Other University and medical libraries, Federation Institutes and various Pharmaceutical companies
Government libraries.
● Dirrect (but expensive) computer searches:
"Medline" from Washington, U.S.A.
Medusa (experimental only, in 1974) from Newcastle, England.
● Even Public Libraries in most big towns carry a certain number of books which may be well-worth consulting when reading "around" the proposed research project. The Westminster Public Library, in St. Marylebone, has a large collection of medical books and journals.
● L. T. Morton has published a book, *Use of Medical Literature* (1974. Butterworths, London) which deals specifically with information sources for research and development. With chapters by different authors on separate specialties it provides in one volume valuable references to almost every kind of medical research (Chapter 14 on "Clinical Medicine" by Miss E. M. Read gives a detailed description of where to look, for a start; other chapters are for specialised interests).
● Finally, always remember that personal observations are the most important source of information; if the printed word conflicts with what you have seen, then don't believe it and say so in public when the evidence becomes available.

SPEAKING ABOUT RESEARCH

Almost as soon as you have started your research project some fool will ask you to talk about it, informally and not as a defined lecture. What to do? Reject the invitation if it gives you little time to compose a coherent story – at the same time negotiate a future date. It is probably wise not to refuse outright but to take the opportunity to make use of the earliest occasion in three ways:

- to impart information. Perhaps only to outline what you intend to do, perhaps to describe the results of a pilot study, perhaps to provide early results (a kind of "in progress" report).
- to invite criticism at an early stage, certainly before the proposed paper is anywhere near ready for publication.
- to obtain new ideas from the audience, which may help you later.

If you don't aim for these three objectives then you will be wasting your own time and that of the audience. All this means that your short dissertation must be contained well within the time allotted and an effort made to induce the audience to participate; the easiest way is to ask them questions with a distinct air of humility.

It is useful to remember that audiovisual aids may help enormously. For instance the:

Blackboard Overhead Projector Lantern Slides Films T.V. Audio-tape	may be used to: 1. Support the spoken word 2. Provide a visual experience 3. Aid visual imagination.

If lantern slides are to be shown, for convenience especially when speaking away from home, then remember also that they can be used in three different ways:

- as a peg to hang the lecture on: often by word slides or slightly irrelevant pretty pictures.
- to support the spoken word, concept or idea, by reinforcement in a visual way.
- to show something which the audience, perhaps, has never seen before. A slide then can act for the audience and lecturer:

a. as an aide memoire, and
b. to reinforce one point at a time, and one point only, in simple terms. But remember the 2 × 2 law: "The intellectual content of a lecture is inversely proportional to the number of slides shown"; also that the art of discovery lies in stating the obvious first. It is therefore worthwhile being controversial while confessing mistakes in humble language, but also to make facts so obvious that statistics are not necessary at this stage. Intellectual honesty is so important that what you try to put over to your audience should be truthful, original and clear; obvious enthusiasm for your point of view is out of place and may arouse resentment in others who might have thought of the idea first, but didn't. Finally, premature disclosures may be damaging because there is no secrecy in what you say.

STATISTICIAN

New techniques are being evolved by statisticians to present data in the most acceptable form and to discover new relationships between variables in an investigation. Some of these you may never have heard of and so may be at a disadvantage when publishing. So much research relies on good statistical evidence for support so why not a partnership from the beginning?

A statistician should always be consulted if the work is likely to involve measurements. The actual computations may be simple and not require his services (and he may tell you so), but he can check your results, act as a mentor and may give useful suggestions for future work. When the statistician manages a fair proportion of the analysis of the investigation he should always be invited to become an author of the subsequent publication; an acknowledgement at the end of the paper is often insufficient and plainly unjust.

STATISTICS

"There are lies, damned lies and statistics", the words of a politician who would have used the last if they suited his purpose anyway. This is nonsense. The word "statistics" is used in two senses; firstly, in the plural it refers to the collection and analysis of data (descriptive statistics) and secondly, in the singular curiously enough, to statistical theory and the methods whereby data are analysed. The concept of comparison is, of course, at the centre of much human activity; it is the basis of statistics as much as it is of the controlled clinical trial.

Statistics is a body of methods for making wise decisions in the face of uncertainty; it is not a body of substantive knowledge but a method for obtaining knowledge. These two statements may come as a shock to some who believe that statistics can prove anything; they can if the information is dud. Moreover, if statistics are required only to show the results you want, then probably the work is neither very exciting nor important.

Lord Kelvin once wrote: "When you can measure what you are speaking about and express it in numbers, you know something about it: when you cannot measure it, cannot express it in numbers, your knowledge is of a meagre and unsatisfactory kind. It may be the beginning of knowledge, but you have scarcely in your thought advanced to the stage of a science" (or technology).

Science started with the certainty of a determined universe, with the idea of probability as a man-made convenience. We have now reached the stage in which probability reigns and the idea of cause and effect is a man-made convenience. We can thus concur with the sentiments of Sir Francis Bacon: "If we begin with certainties, we shall end in doubts; but if we begin with doubts and are patient with them, we shall end in certainties".

The object of statistics is to measure facts and probabilities so as to draw out their inner meanings, but before one can say what a set of data "means", it is first necessary to be certain that one knows what it "says". The claim that figures speak for themselves is often quite unjustified; it may well be necessary to translate them into a language understandable to others or so to rearrange them that they speak more loudly. But, again, in order to translate from one language to another it is essential that the translator thoroughly understands the meaning of the words used in the original language.

This is a basic requirement in statistics. The variables measured must be clearly and precisely defined. Obviously one cannot measure anything unless one clearly understands its limits and those who have not made the measurements must, before they proclaim the derived measurements, ensure that they also have a precise image of the thing measured.

It is a common misconception that statistics are available to extract information after the research has been done. On the contrary, statistics should:

- impose discipline in planning a project by defining what variables may be dependent or independent,
- avoid distortion of results from other variables,
- help to forecast how many measurements may be necessary before concluding the experiment,
- allow "significance" to be calculated mathematically, in other words, to avoid disclosing a result which might have occurred by chance,
- not be used to infer cause and effect relationships, even though there may be a high correlation between the two.

Every researcher should learn and understand the basic knowledge required, so that he may realise the importance of the subject in his day-to-day work and the limitations of this form of analysis. Statistics are valuable in testing an hypothesis, not initiating a discovery, but even then the results must be viewed in the bright light of common sense; the failure to do so only lays the foundation of a publication by your critics, so don't give them the chance. To understand the subject, start with some knowledge of how statistics may be misused, by reading the book by D. Huff, *How to lie with statistics* (1958 Gollancz, London) and W. J. Reichmann's *Use and Abuse of Statistics* (1961 Methuen, London).

There are many good books on statistics and it is not intended here to write another. As with all research, one learns by doing, so too with statistics. Hence working out the means, standard deviations and performing a test of significance on your own data will teach you more than reading about methods. Three books are highly recommended:

- *Facts from figures* by M. J. Moroney (1965 Penguin books, London) because it deals with the subject in a general way and is easy to read.
- *Statistics in small doses* by W. M. Castle (1972 Churchill, London) because it provides just that.
- *Principles of medical statistics* by Sir A. Bradford Hill (1971, 9th Edition, Lancet, London) because it is eminently readable and has survived the passage of time.

When these three have been digested, there are numerous specialty books which may have to be read later. But by that stage you should have a friend in the field, a statistician. You will also have come across and perhaps obtained a copy of Documenta Geigy Scientific Tables, a book of reference data invaluable to anyone who knows where to have it to hand – on your desk.

SUCCESS

There are no technical rules for success, but the conditions for assessing and planning a research project may decide the result; there are 12 things to keep in mind:

- The project must be made effective, that is, the end product must be visualised early on. It is no use beginning without an hypothesis.
- Its potential must be identified at the start.
- It must be made into an investigation for a different future. Few projects will succeed if they concentrate on the past: history sees to that, not research. The conclusions from the work must be for action by yourself and others.
- The only resource for success is knowledge: money and physical equipment do not confer any particular distinction on the project even though they may be essential requirements. Knowledge, however, is not enough; one has to be a swordsman and a gownsman.
- Results on the whole are obtained by exploiting opportunities and not by solving problems; heresy to many, but in the main true.
- Hence resources must be allocated to opportunities.
- Successful research results are earned by leadership, not by mere competence (Drucker again!).
- Any position of leadership in a field is transitory and confers only a temporary advantage. What one man has done, another can always do again, often better.
- Once the project is underway it is already becoming obsolete. Hence, success depends very much on planning the next stage. For this reason allocation of resources must be constantly shifting. A deliberate re-appraisal and redirection of effort must be carried out.
- The same few men in a laboratory are apt to produce most of the important innovations and advances. They also produce the

most results from the resources.

- Concentration is the real key to successful results. This means that diversified efforts produce less, or even nothing at all. It may not be possible in a unit to combine teaching, clinical practice, a community service and research. A decision will then have to be made which item should be abandoned. Most people know this to be true, but find it emotionally hard to accept. Yet, in the final analysis it has to be done.
- The strict use of the scientific method approach in research is not just important, it is vital.

Success in research, as in life, is accomplishing what you set out to do. It may be major or minor. Unfortunately, success is also judged by others – by the number of publications, contributions to a new knowledge, recognition of expertise as an authority in your subject. Success depends upon finding a solution to the problem proposed. When a solution is difficult to find, the reasons commonly are:

- The problem has been ill-defined. So, it has to be approached from a different and entirely new angle. In this respect a project postponed is a project abandoned: rarely can it be resurrected effectively, so cut it out. You just have to be ruthless with yourself as in politics.
- The information collected is wrong, because it is incomplete, irrelevant or untrue.
- The action taken is irrelevant even though the intention and logic seemed, originally, to be good.

A good project is essential, but courage rather than convenience is the motto to adopt. Finally, successful research *demands*:

working time, for doing things
leisure time, for planning
money
thought and originality
concentration
ability
interest and enthusiasm

If you have none of these to give, then don't even start.

TALENT

To a certain degree talent is innate, but there is no doubt that it can be developed by good practice. This means that juniors must be taught (and this takes time) by the best in the team. A good precept is "See one, do one, teach one". Natural talent must be recognised early on, fostered and nurtured.

TEACHING RESEARCH

It is the duty of an experienced researcher to teach the novice. But how? Techniques and the basic concepts will be imparted and much gained by absorption anyway, but there are 7 special features which must be put over:

- Self-confidence. If a junior is allowed to do one thing well, he will not be afraid to attempt another. He will then become what research needs most, a problem-solver. To develop self-confidence the junior must be provided with opportunity and encouragement. One cynic said that "confidence will get you nowhere in research, only hard work". Maybe, but self-confidence is something apart.
- Enthusiasm for the project will probably come naturally. Pavlov said: "Firstly gradualness ("festina lente" was said 2,000 years earlier) secondly modesty, thirdly passion". Enthusiasm fosters the desire to work hard.
- Compassion for others, if only to keep the lab. man human.
- Respect for truth, honesty, moral values and quality. The junior has to develop a quiet conviction of the importance of all four of these. Personal honesty is of the utmost importance and should be pursued pitously and mercilessly, for without integrity what is left?
- Resilience, that is the ability to cope with change, to bounce back. The ability to laugh at oneself is invaluable, for humour can make a grim situation more palatable.
- Hope, that the problem *will* be solved, that questions *do* have answers, that challenges *can* be met.
- Courage, to stand up to criticism, to defend what he believes to be right. Later in life he will have to defend his new ideas which may conflict badly with those of his superior; he may have to present ideas for discussion and argue his case, or push for research grant funds. The one quality which will stand him in good stead is another four-letter word, "guts".

Finally, advice for the teacher of research: if you preach, practice; if you don't practice, don't preach. Unfortunately, university promotions are based primarily on scholarship (as indicated by the number of publications) and on judgement of research competence by colleagues and not on teaching ability. Hence, many of the most dexterous are so often ineffective teachers.

TEAM

Researching on your own is fun. Unfortunately, it doesn't last long. Quite soon others become involved, by expert advice, the helping hand, special favours, technical help, and eventually a partner. Even with two of you, there is a team: but it rarely stops there, the team grows of necessity. So now we have to consider seriously how to build a team. This is not a matter of gradual accretion, like barnacles on a ship, but of picking the expertise and the quality you want, of finding people who will further the work. A good research team always starts small and ends big.

An effective team, rather like a happy family, will have 11 special features:

- There will be an atmosphere, which can be sensed, of informality, relaxed and comfortable people: all are interested and involved.
- There is a lot of discussion in which virtually everyone participates, and it will usually be pertinent to the research project.
- The research task will be clearly understood by all and accepted as reasonable. Moreover, free discussion will continue until the objective has been defined and all are committed to it.
- Members of the team listen to each other and no one is afraid to appear foolish.
- There will be disagreements because the group does not try to avoid conflict. There is no tyranny of the minority, for reasoned arguments are examined impartially. The team tries to resolve them.
- Decisions will be reached by a consensus opinion where everyone is willing to go along with the general agreement. On the next occasion, some members will unblushingly reverse their approval, if personal observations have convinced them otherwise.
- Criticism will be frank and constructive.
- Everyone will feel free to express his own ideas.

● When action has to be taken there will be clear assignments for each individual and willingly acceptance.
● The leader of the team does not dominate in discussion and nor do members of the team defer unduly to him. There is no evidence of a struggle for power.
● The team will be self-conscious of its own work and often will stop to see how well it is doing. Colleagues are individualists and are not prepared to have this quality submerged. All confidently expect to use their own personal resources to complete a genuinely creative job.

Compare this picture to the ineffectual, incompetent, unhappy, poor team:

1. The atmosphere is of indifference, tensions, or boredom.
2. Only a few people dominate the discussions, and they are often off the point.
3. No one listens to another, ideas are ignored or are overridden.
5. Disagreements are not resolved, but suppressed so that open warfare and a majority vote is the final solution.
6. Action may be taken prematurely, before the central issues have been examined in detail. A simple majority decision may be considered sufficient for subsequent action, the minority of dissenters being expected to comply.
7. Decisions often are not clear and nobody knows who will do what.
8. The leadership position is clear for he always sits at the head of the table, whether a weak or strong chairman.
9. Criticism is always embarrassing and leads to tensions. Personal hostilities are evident but often hidden. New ideas receive destructive criticism and so no one is prepared to stick his neck out.
10. There is no discussion of any individual's performance, and therefore no place for praise.

This all sounds rather depressing, but there is a special potential in having a team:

● A group target offers advantages that cannot be achieved by an individual alone.
● A team offers an excellent environment for the individual worker to develop his own talents and to appreciate the need for

collaboration. There is good training in social interaction and in solving problems.

- Significant measurement of performance can be developed for the group which cannot apply to a single individual.
- Unity of purpose becomes possible in a well knit team. A sense of purpose and direction is important in Government, research team, the Army – call it what you will but things get done by all pulling in the same direction (even if it's the wrong direction) whereas nothing is accomplished by pulling against one another. Indeed, this high-lights the principle of integration which depends on: consultation; argument; persuasion; recognition.

Integration means working together for the success of the enterprise, so that all may share in the final reward. The implicit assumption, that working together always means absolute direction from above is quite unjustified (see McGregor's theorems). Every team benefits from having a code of practice: it should be implicit and not recorded in words (that's a "job description" anyway); it remains the unwritten law, but certainly understood by all. In a good team everyone knows how to do everything and can "double-up" for an absent member even if not quite as expertly. If you don't know what the team spirit really means then join a cargo steamer for a long voyage. Each man is engaged to do a specific duty; if he can't, because of illness or injury, then someone else has to work doubly hard. In the vast expanse of the ocean there is no relief substitute to be brought out of cold storage to fill the gap. As a result, seamen return to duty as soon as they can (sometimes sooner). This attitude, not necessarily condoned, is seen in armies of high morale, in times of great unemployment and in national disasters, where it earns universal praise; it should develop quite naturally in a research team, as a sense of duty to others.

The Peter Principle states that "the cream rises to the top until it sours". Unfortunately, so does the scum, hence it is of more than passing importance to know something of the mixture below. Maybe this is the essence of building a good team of competent industrious people. The work-shy have a habit of not changing character even when opportunities are presented to them.

TECHNOLOGY

Medicine is not a science, as so many people think. It is a technology. Mathematics, physics, chemistry all deal with exactitudes: they are sciences. Medicine, often called a biological science (a contradiction in terms) deals with variables. A technology may apply the principles of science, but the important word is "application"; it is focused on use.

Technology is how things are done or made: it has to do with work, the human activity whereby man can alter his environment. This is not to say that there are other factors playing a part in technology (Drucker, 1972) – the legal system, economy, political institutions, social values, philosophical abstractions, religious beliefs, scientific knowledge, medical practice. They all have a shifting relationship as Drucker points out, and "all of them are part of technology in one way or another, at one time or another". Technology must be considered as a system, that is, a collection of inter-related, interacting activities. Technology is ideas in action – and so worthy of medicine by any standard. Scientific discovery has always been measured by what it adds to our understanding of natural phenomena. Invention, however, has quite a different measure – what it does to allow us to carry out a specific task. Technological invention then has quite a different approach, to change man's environment. This is the real impact of technology and the sole reason why Western technology is in such demand. Medical research has only one view in mind – the betterment of a patient.

Drucker (1972) is quite adamant. The scientific revolution may have provided the "scientific method" for general use, but the "Technological Revolution" is of more recent date. He believes (and I have to agree) that the major changes in agriculture, engineering and medicine occurred between 1750–1850, a process owing little to the new knowledge of contemporary science. Drucker goes further and argues that technology dragged science with it and made the latter respectable. In every technology "the rule of thumb was far ahead of science". It is difficult to argue otherwise. Knowledge had changed its meaning from "understanding" to "control".

Medicine has been fortunate in that practice and theory have been taught for over 2,000 years. Yet the observations and applications of Jenner and Harvey had no relation to science, only to application: indeed, a century was to pass before a scientific explanation became available. Observation, not scientific knowledge, confirmed their work. And so it is to this day, where most research is based on observations, that explanations follow in their wake. This is not to say that scientific knowledge is not being translated into technology (it is!) and vice versa. Technology is focused on use (Drucker again) whereas science is more concerned with understanding. Both are essential, but they do not march together. Medicine has a foot in both camps: a fortunate position.

TESTIMONIALS

Sooner or later, a member of your team will find promotion elsewhere. If there are good family relationships within the team, he will quickly ask for a reference. Testimonials are mostly pretty useless, but time will have to be spent in composing a eulogy.

Testimonials are two-way affairs. When you come to replace an important assistant, he too will bring letters of praise. Why not pick up the telephone, speak to the candidate's previous employers one by one? Identify yourself, explain that you have permission and ask for verification of the information given to you: check his previous position, salary, responsibilities and reasons for leaving. If all these are favourable then ask about loyalty, leadership qualities, ability to get on with others and to work hard, absenteeism and personality. In five minutes you will have learnt more than from reading ten pages of commendation. This is using the telephone intelligently and usefully: the tone of voice at the other end will tell you more than any letter ever can.

We buy brains whenever anyone is taken into a project, for that is the only commodity that will produce results; the price is negotiable but often not commensurate with the final product. Even so, you may have to state clear incentives (such as a thesis, promotion or publications) when you find the right person. Attitudes and temperament are important qualities for the man to fit into your team, but the lack of enterprise and imagination in the selected one may remain as evidence of poor judgement for a long, long, time.

THESIS INDUSTRY

Most research ends up as a paper fit for publication. It reports something new in intelligible language. Some are longer than others, but it is a mistake to think that a paper can be expanded to a thesis – sure it can, but failure is courted. Of course, it is nice to obtain a coverted university degree in your subject; If you really want one, then the research project must be worthy of the prize (neither a review of previous work nor a collection of patients, but something new and original).

Writing a thesis is a major undertaking. To be accepted it must report new work and comply with the University regulations. The first thing to do therefore is to request the necessary regulations, to read them carefully, and, if in any doubt about their interpretation, to consult an official for clarification. Many universities insist on the candidate having a supervisor: he should be chosen with care so that he knows your subject sufficiently to give good service and has time enough to devote his whole attention to regular consultations. A thesis may fail simply because the supervisor is inadequate, so look at his track record.

To be successful is a different matter. Without self-discipline and hard thinking it cannot be done! Without enthusiasm it becomes a dull chore, which shows in the writing. There must be sufficient time to carry the project through, perhaps three years of part-time effort or one year of full-time work. Some take longer, but if after five years the thesis has not been completed it probably should be abandoned. The thesis is an educational tool. It is the result of one individual's work and should present a formal statement as an hypothesis, preferably with more than one approach to the topic. It should present all the data obtained in the study. It also offers the opportunity for an extended and argumentative discussion; it is meant to be a scholarly dissertation. A thesis, however, does not have to be thick to be erudite, and in my view, most are too long and make tedious reading; size and quality do not go hand in hand.

It is presumed that preliminary explorative work has already been done on the proposed subject, and that a protocol of the definitive project has been written. The candidate must choose his own subject; it should be one in which he is already interested, and better still, one for which he has already carried out a great deal of work. He should have had some experience of writing and of the principles of research. In choosing a subject one has to be especially wary of the ideas which seem the most self-evident and the most obvious; they are current and we have heard them accepted as truisms from early days. Yet it is exactly these ideas which should be examined more closely as the basis for an exciting thesis.

It helps to remember that there are 6 stages in the preparation of a thesis:

● *Early planning*
Right at the beginning a précis of the work should be written on two sheets of paper. This brief outline may be submitted with the application to the university, for approval to proceed. Because the thesis will be a study in depth, the title demands particular attention; it should be well defined and well confined. You must be clear about the problem under discussion and why you selected this one for study; it is that much harder if you have no imagination.

A great deal of reading will have to be done, and relevant findings should be made on 15 × 10 cm. record cards, together with the exact reference. Reading should be selective and perceptive. It is a truism that good writing follows good reading and the wider the variety, the wider the general approach to the subject. The candidate is expected to know all the resources of a library and the best guide is to enlist the help of the medical librarian. It is usually impossible to review the literature completely so you have to use judgement. In the same way, if statistics will feature in the finished work, (as they should) the advice of the statistician must be sought at an early stage.

Finally, ideas should be recorded on paper within the recognised structure for the finished work. Some universities lay down precise details for the lay out, binding and the general format. To some extent the structure depends on the work reported and it is advisable to look over successful theses of similar nature. In any case, many subheadings will be needed to make it readable. Few theses flow like a well written novel.

A possible analytical outline might be:

1. Title page, with name, degrees and date of submission.
2. The curriculum vitae of the candidate.
3. Abstract of the work, of about 50 words or so.
4. Summary of the main findings perhaps 400–500 words.
5. Survey of the relevant published work, described in depth. This should not be a chronological list from earliest times, but a well-considered account strictly limited by the title. This introduction should be no more than one quarter of the whole thesis.
6. Materials and methods.
7. Patients studied.
8. Results, in chronological order or in order of complexity. Tables, graphs, diagrams, will all be included in this section which should form the major portion of the script.
9. Discussion, that is what you set out to do and what was found. Sub-headings will be essential when changing from one topic to the next. If you mix your own data with other published work, utter confusion will follow.
10. Conclusion.
11. References.
12. Acknowledgements.
13. Appendices for detailed data, which if retained within the text, would impede the flow and rhythm of the script. These may be tables, case reports, or special analytical techniques.

An alternative structure more suited to work which has included several experiments of a different nature would be:

1–4. as before.
5. General introduction.
6. Experiment I
   a. Materials and methods.
   b. Results.
   c. Discussion and Conclusions.
7. Experiment II
   a. Material and methods.
   b. Results.
   c. Discussion and conclusions.

This sub-structure can be repeated as often as required.

8. General discussion of results.

10–12. as before.

Each section heading should be written on a separate sheet of paper, to receive notes of data and ideas. During this period of assimilation it is advisable to purchase a box file 27 × 36 × 8 cm. into which are placed all scraps of paper with bright ideas, library reference cards, cuttings from journals, reprints, preliminary data, and notes recording conversations with colleagues.

- *Definitive Planning*

One day when you have a couple of hours of quiet, empty all the contents of the box file on to a large table. Now is the time to sort over what you have acquired and to decide whether the thesis is possible, or even likely, within the time available.

The definitive research may have only just begun. Even so it should be clear by now what aspects of the subject will be the most demanding in time and effort.

- *Preliminary Writing*

At this stage the project should be going well. Now is the time to start getting fit to write well. Start writing the introduction, material and methods, some results and part of the discussion. There will be gaps of course for not all the data will yet be available. During this time, leisure hours should be spent in reading good English, the influence of which will be noted in better writing later. Good reading is intellectual nourishment, as important to the candidate in producing a scholarly work as the data on which the thesis is based. Everyone writing a thesis must cultivate some literary values, an important part of the exercise. The manuscript should not be typed at this stage but written in long-hand leaving plenty of space for corrections.

- *Handling the Data*

As each set of experiments is completed, tables should be constructed. Examination of the results in this form will indicate to the candidate what new work may have to be done to support the original hypothesis or to substantiate the likely conclusions. A thesis which proves everything and explains nothing is hardly meritorious; critical analysis in interpretation is mandatory.

- *Writing*

As soon as the work has been completed, it is wise to take one or two weeks leave to write and assemble the whole thesis in one session. By doing so, continuity of thought, unity of composition, flow and readability, are likely to be assured. This stage is undoubtedly hard work and the most difficult physical act to undertake. But it has to be done. The thesis has now been written, an ample reward for the effort.

● *Revision, Rewriting, Binding, Submission*

After the torrid atmosphere of the previous stage, revision is almost an anticlimax. The wise man will wait at least a month before revising his script, so that it can be examined as though it were a new work. Revision and rewriting can be taken at a gentle pace. Most theses can be improved by skilful editing: many are rejected because they are badly written. They do not do justice to the facts they contain.

The work should be read to answer three questions:

1. Is the logic correct?
2. If the logic is sound, are the basic assumptions at fault?
3. Are the conclusions justified from the evidence presented?

If the answers are "yes", apply the rules of good English, examining each paragraph at first and then sentence by sentence. Prune relentlessly, until the manuscript is to your entire satisfaction. Finally, have it typed and read this version for errors. The value of the thesis lies not so much in the quality of the work done but rather in the personal discipline which has been acquired.

In all instances, three persons should be allowed to read the final script:

● The Head of your Department, who may be one of the university's assessors. It is embarrassing for both of you if he has had no prior knowledge of the work.

● Your supervisor, who may not have had the opportunity to see the completed work in final form.

● A friend, or colleague, or if possible your wife, to assess the literary quality and clarity of the writing. Even if their opinions lead to further revisions and corrections, the gain is usually rewarding.

A good rule is not to publish any detail from the thesis before it is submitted to the university. Once the degree has been awarded, parts of the work can be submitted to the journal of choice, but will require rewriting. If the work is so important that some aspect should be published early, to establish the claim of priority, then it is essential that the candidate's name be first when there are multiple authors.

THINK TANKS

"Think-tanks" and "brainstorming", two ideas which have travelled across the Atlantic, may be successful in industry: they rarely are in medicine. Leave these activities to the clever ones with large and expensive research units, and be content with a little peace and quiet to think on your own. When your thoughts have clarified, discuss them with a colleague by all means, but don't be put off too easily.

TIME

Time cannot be bought or sold. It is totally inelastic; a commodity which can be used, or thrown away as though it had no value. Time is precious and valuable in any research; it can be conserved but not preserved. Waste an hour and it never returns. This is not to say that looking out of the window while planning the next project is wrong or wasteful; rather, an ill-conceived plan which leads nowhere is time wasted.

Time is so precious that you will have to decide how you spend it, like currency. To cut down on social obligations? Less on reading (but can you afford to)? In the last resort, the wise one will sit down and analyse how the 24 hours of each day has been spent. He will then know where his money has gone and seek to save. In surgery, it is possible to conserve 80% of time and 50% of effort by thinking ahead, by training everyone to know what has to be done and when. How do I know? Quite simply by having a stopwatch and timing a standard procedure: that's research, not boasting. The problems then are how to find out where time goes and how it can best be spent.

In all research the most scarce resource is not money or ideas (even though there would be no research without them), but time. Time is the essence of almost everything you do. Plans may look good on paper but without time they cannot turn into achievement, and that's what we're after. Hence, the effective leader should not start with tasks but with time (Townsend's time-sharing principle). He thus has to start by deciding where his time should be spent and then examine how it actually is spent; he must retain control of his own time.

However much the demand there is a limit to the day's work. Time is irreplaceable, yet all research takes time. Time can be conserved by:

- Cutting out unproductive and wasteful tasks. This may mean avoiding committee meetings, conferences, travel, official dinners, superficial entertaining, the lot. You really do have to be ruthless. One solution is to delegate and foot the bill if necessary.

- If time must be devoted in a large chunk, say, 2–4 hours to write a report, to plan details of a project, to write up results, then it should be so apportioned; to arrange 4 half-hour periods instead is neither effective nor wise (nothing will get done that way anyhow). When the script is finished, smaller periods of time for corrections and alterations are perfectly in order. Rewriting, sentence by sentence is a fairly relaxed past-time: the main issue is not.
- Time for experiment just has to be in one block, a continuous effort; it cannot be split up.
- People are time-consumers and most of them are time-wasters, so be ruthless. If the interview is important, then make time and make sure the job is done: be sure, too, that this interview does not lead to further fruitless discussion.
- If a member of the team has done well, make time to discuss his results in detail. Learn all you can of his work, ask pertinent questions, make notes, direct the next stage, use the time for profit. If you cannot spend time on success, then resign from the leadership position: it's as important as that!
- When time is devoted to a vital issue, make sure you can spend it in one lump. Thus, when interviewing a new applicant for your team, and that is as important as anything, give the impression that time is unimportant. This is not to say that you do not have to get things done fast, you do, but sleight of hand does have advantages.
- Always make time for innovations, new ideas, new concepts. Hence you will have to manage your own time irrespective of routine.
- Eliminate things which need not be done at all. The advantages of a "pending" tray are obvious.
- Decide who can do things just as well or better than you, in their time, and delegate.
- Don't waste the time of other people by pottering around when work is going well. Poor management in a research team wastes everyone's time. A plethora of useless meetings must be avoided; one meets or one works, not both.
- Ask if over-staffing is a cause of time-wasting. If an expert is

required infrequently it is better to hire him on a part-time basis; he will then keep away until called.

- Will your time make a significant contribution to the research project? If the answer is "no", then forgo the honour.
- Use a secretary to take messages and answer routine questions.

TITLES

Often the best research is done by one man and a boy – working hard and together. When a team effort is required it is absolutely vital to have too few people rather than too many. This is the time to designate titles. Call him "technician-in-charge" and his shoulders go back to do better; the pay is the same but the status quite different. An example of a distinction without a difference.

Remember to say, "Research Fellow", not research assistant, and "Public Relations Officer" for the secretary. The word "Deputy" is another useful office, but try to delegate something. Better still, use the titles in publications when these valuable assistants, whom you could not do without, are co-authors.

TOLERANCE

Somerset Maugham in his book, *The Summing Up* (1952) wrote that "tolerance is another name for indifference"; the dictionary defines it as "forbearance" which is not quite the same. Tolerance in research does not mean mediocracy but rather:

- having to admit that you might be wrong
- recognising that you could have done the other man's work so much better, but didn't
- realising the necessity for comfort when things go wrong, the misery of mere mortals
- forgiving others whose publication anteceded your own, even though you may have started the work earlier (but failed to complete on time)
- accepting that others too make mistakes.

In the end it's a kind of immunological mechanism that improves with time.

TRACK RECORD

Everyone engaged in research should examine his own performance from time to time. This "track record" can be an indicator of how well you are doing and what sort of status you deserve among colleagues. It can be a painful exercise, but it is necessary.

The number of papers published divided by the number submitted should not fall below 80% and the number of projects started divided by the number completed should come to about the same figure, or better. If either falls below this arbitrary level then it is time to rethink, or the time to get out. With such a degree of uncertainty it would be unfair, for instance, to supervise the thesis of a junior who will be relying heavily on expert direction for success. If judgement and interpretation are good your track record will be likewise; when imagination dies, so does your reputation.

WHAT RESEARCH?

It is possible to discern certain groupings in the areas of research. These apply whether the research is done by doctors, nurses, biochemists, veterinary surgeons, or the para-medical services (such as physiotherapy, radiography, medical illustration and a host of others).

Anyone looking for a problem to investigate should examine this simple classification for help and then think hard.

- Diseases: such as diabetes, to look for new syndromes, new features, morbidity, mortality.
- Techniques: of investigation or of treatment, whether they be medical or surgical; in other words, methodology.
- Management: of services, such as how clinics are manned and their efficiency or cost, time spent in hospital for investigation or treatment.
- Education: either in methods or effectiveness, both for staff and patients. Each of these four may be subdivided further into:

1. Descriptive, which may be retrospective (to find a good idea) or prospective (putting the idea to the test).
2. Manipulative, which is always prospective.

Retrospective research is usually dangerous (and very few papers are accepted for publication by reputable journals), if only because there is usually a hidden bias in the population being surveyed and (frequently) missing data. The value of retrospective research lies in providing new ideas which may be included in the final writing, but they are relatively unimportant.

Scientific research works best when all variables are held constant, except the one you are testing. Even so, prospective research is expensive in time, effort and resources. Research ideas are two-a-penny for only a moment's thought; good projects are more difficult to find and are successful only when paid for in sweat, toil and tears; even then they may not be worth the effort.

So why work? If you still have doubts, some reasons are provided under "Why?".

Here is a list of subjects which might be tackled in a medical specialty; it is applicable to many others and is by no means exhaustive. It does, however, concern the man who says, "What can I do research on?"; anyone who asks this question has not thought about all the possibilities with any sort of imagination.

1. Administration:
   How to deal with the current waiting list.
   How to improve the turn-over.
2. Traditional procedures:
   Where are the supporting statistics?
   Can they be simplified?
3. New procedures for common conditions.
4. New knowledge:
   How do things work?
   Why do things work?
5. Divisions into clinical and experimental:
   Can they be combined?
   Is the division largely artificial in modern medicine?

WHEN TO DO RESEARCH?

The short answer is "now" (Fiat Nunc in Latin). There is widely held belief that, for most people, originality declines at an early age. Maybe. So why not start young, as soon as you can read and write and use your hands?

Beveridge (1951) pointed out that research is one of those highly complex and subtle activities that usually remain unformulated in the minds of those who practise: "this is probably why most scientists think that it is not possible to give any formal instruction in how to do research". This view was not shared by him and the present volume represents an attempt, not to teach research, but to provide information about important aspects and the trivia which only experience endows. With such help research investigations can begin right now. There is only one motto: hats off to the past, coats off for the future. It takes a lot of learning and a lot of hard work.

Methods for the collection of new knowledge, for that is what research is all about, vary. Technique can be learned fairly quickly; the essential philosophy and practical organisation take longer. Personal experience is a wasteful method of acquiring this knowledge and hence the best advice is to learn how to do things correctly by joining a competent team. The more practical matter of making sure that you can get on with the job undisturbed requires some organisation. One way is to have another person take all your calls with the promise that you will phone back between certain hours. Let the caller know that you are in the lab. (some kudos for research there), but do deal with problems when you say you will.

WHERE TO DO RESEARCH

Some people who desire to carry out research simply do not know where they can do it. The quick answer is, "in your own office, or at home". But research usually demands special facilities so that really it cannot be done anywhere. There are four ways of finding out where to work:

1. Ask around at your own institution to discover who is doing research similar to that proposed. Bench space may be at a premium and at first an offer of another pair of hands, by helping with someone else's project, may have to suffice.
2. There are several books about Scholarships (which see) and perusal of these will tell you of likely laboratories where you would be welcome, for many scholarships are linked to certain institutions. The advertisement pages of the weekly British Medical Journal and the Lancet carry notices of possible openings.
3. There are also books such as *Scientific Research in British Universities and Colleges* (1973–74 Volume II. Biological Sciences) issued by the Department of Education and Science of Government which give a guide to the subjects under investigation at various institutions, a weighty volume of 674 pages.
4. Annual reports from the Medical Research Council, the Wellcome Trust, the Nuffield Foundation all give useful information and those wishing to work outside their own country should consult the *Scholarships Guide for Commonwealth Postgraduate Students 1975–77* (Association of Commonwealth Universities, 36 Gordon Square, London).

WHY DO RESEARCH?

Research is the quest for new knowledge; the purely physical activity of experimentation is only a means for supporting or rejecting an hypothesis. Knowledge, however, is a perishable commodity: it has to be reaffirmed, relearned, repractised all the time. Knowledge can become obsolete too – hence, there must be continuing research asking, "How can it be improved?"

A researcher is often overworked, usually undervalued, never overpaid, so why do research? When you become aware of another person's need and do something about it, you begin to understand what research is. If you do it – and it may be quite undramatic – you will make the world a little richer, a little better. It's worth a try. There are 12 reasons:

1. Because it's there, a simple answer but vital to progress.
2. A love of gadgets and gadgetry.
3. Idealism: its not enough to be concerned, so do something about it. Research in medicine, unlike that in industry, is for the general good not just for the individual and there are no secrets.
4. Curiosity about the world around us: an essential ingredient.
5. Challenge. There is an old village saying that if you don't buy a raffle ticket, you can't win a prize.
6. Discipline and self-discipline, which is always good for the soul.
7. To obtain publications (the end product of all research) for personal advancement in your profession; a just reward for effort and not to be decried.
8. Imitation of others.
9. Prestige, but a rather ephemeral currency.
10. To obtain grants.
11. Sheer fun – surely a vital ingredient for life and often the only thing remembered years later.
12. Gamesmanship: Research is a game; it can be played hard

or easy, professionals against amateurs and very often the amateurs have a habit of winning, which makes it open to all. Research is interpreted as "them" against "us" but in the end, all are "us".

These 12 reasons are not in order of merit and there are others, anyway. On the whole, I would put the sheer fun of discovery first.

Finally, of course, research does pay. In industry, it leads to new methods, new products and sometimes higher profits; in medicine new methods may shorten diseases, improve prognoses, reduce morbidity and mortality.

WRITING

Write so that others may learn. Do not write for your own glorification or to expose your erudition – both are ephemeral. They will come, if deserved, by the quality of work done and not by the quantity of words used to express it. The difference between a well-written paper and a bad or indifferent one is that the good can be read more quickly and its meaning grasped more readily. So think always of the reader. If your message is misunderstood you have only yourself to blame.

Bland and meaningless statements, popular in politics and diplomacy, have no place in scientific writing. So don't write them. There are several good books for the embryo author: consult them, especially the small volume by Calnan and Barabas (1973). Also read Asher's collection for pleasure and profit (especially Chapter 4, The Six Honest Serving Men).

In general there are three separate sections to a paper:

- the beginning
- the main data
- the end.

When the work comes to be written up, 10 separate sheets of paper should be obtained and these titles put at the top:

1. Title and authors
2. Introduction
3. Materials
4. Methods
5. Results
6. Discussion
7. Conclusions
8. Acknowledgements
9. Summary
10. References

Each sheet should accumulate scribbled notes of what must

be included. Writing, however, should not begin in earnest until you have some idea of the results of the research, so make out tables and graphs to direct thought towards the conclusions. Results and conclusions make up the central theme, just as in a film or T.V. play.

Writing requires effort, concentration and peace from surrounding distractions so it is not easy, but if the writing does not match up to the quality of the research, due credit will not be given. Hence avoid long sentences with the vocabulary of a legal brief and write as you would wish to read. Write and rewrite as often as required, until at last you have distilled the pure essence of what you want to say. It's a painful process, especially when a colleague reads it and says "trim it down"; usually if you don't, the editor of the journal will tell you more forcefully.

Of course, if you have a command of language which is staggering, precise and vivid, then ignore all advice. Your papers will be read no matter what the content, but most people do not have these qualities.

ZOMBIES

All through this book I have made the assumption that you, the reader, wished to do research or at least know something about it. But, not everyone will make that assumption and some may even suggest that you should not. They will say that your job is to get on with the work assigned to you, as though this precluded thought. They have no doubts in what they do, so why should you? Their arguments are numerous and persuasive.

For these folk the term "zombie" seems appropriate (the word is now listed in the Addenda to The Shorter Oxford Dictionary). Avoid them or ignore them. To be an observer of life, neither involved nor committed, cannot be you.